Fortress • 53

Defending Space

US Anti-Satellite Warfare and Space Weaponry

Clayton K S Chun • Illustrated by Chris Taylor
Series editors Marcus Cowper and Nikolai Bogdanovic

First published in 2006 by Osprey Publishing
Midland House, West Way, Botley, Oxford OX2 0PH, UK
443 Park Avenue South, New York, NY 10016, USA
E-mail: info@ospreypublishing.com

ISBN 10: 1 84603 039 0
ISBN 13: 978 1 84603 039 0

Design: Ken Vail Graphic Design, Cambridge, UK
Typeset in Monotype Gill Sans and ITC Stone Serif
Maps by the Map Studio Ltd
Index by Alison Worthington
Originated by United Graphics, Singapore
Printed in China through Bookbuilders

06 07 10 9 8 7 6 5 4 3 2 1

A CIP catalog record for this book is available from the British Library.

For a catalog of all books published by Osprey Military and Aviation please contact:

Osprey Direct, c/o Random House Distribution Center, 400 Hahn Road, Westminster, MD 21157
Email: info@ospreydirect.com

Osprey Direct UK, P.O. Box 140, Wellingborough, Northants, NN8 2FA, UK
E-mail: info@ospreydirect.co.uk

www.ospreypublishing.com

Acknowledgements

I would like to express my great thanks to Dr Harry Waldron from the United States Air Force's Space and Missile Systems Center, Los Angeles Air Force Base, California. He helped me secure several photographs for this book. Nikolai Bogdanovic and Marcus Cowper from Ilios Publishing provided superb support and encouragement from the inception to the publication of the book. Finally, my family allowed me the time to work on this book. My wife, Cheryl, and sons, Doug and Ray, sacrificed a lot of "lost" weekends and nights.

Image credits

All photographs that appear in this work are from the United States Government. The majority of the photographs come from the archives of the United States Air Force, Department of Defense and the National Reconnaissance Office.

The Fortress Study Group (FSG)

The object of the FSG is to advance the education of the public in the study of all aspects of fortifications and their armaments, especially works constructed to mount or resist artillery. The FSG holds an annual conference in September over a long weekend with visits and evening lectures, an annual tour abroad lasting about eight days, and an annual Members' Day.
The FSG journal *FORT* is published annually, and its newsletter *Casemate* is published three times a year. Membership is international. For further details, please contact:

The Secretary, c/o 6 Lanark Place, London W9 1BS, UK

Conversion table

1 centimetre (cm) – 0.3937in.
1 metre (m) – 1.0936 yd
1 kilometre (km) – 0.6214 mile
1 hectare (ha) – 2.4711 acres

Contents

Introduction

In the past, most national and military leaders believed space systems and satellites were exotic or in the realm of science fiction. Today, the US chain of command from the President to the newest recruit relies on military space systems to conduct operations. The belief that military satellites are vital to national security and interests is not restricted to the US. China, Russia and other nations realize that the opening of space for military, commercial and civil purposes is a key to the future. Competition to maintain and control space has increased in the past decade and will continue to do so as nations develop and integrate space-based systems into their arsenals.

The US government developed space systems as a means to protect the nation from a nuclear strike. These unheralded sentinels gave national security experts sufficient intelligence data to make crucial decisions that affected the national security strategy for years. Satellites also allowed the country to know instantly if the Soviet Union had launched a nuclear-armed ballistic missile attack. Bomber, ballistic missile and submarine crews could then take actions to retaliate in kind. Assured destruction of an aggressor offered an invaluable deterrent option that helped keep the Cold War from turning hot.

These systems were highly successful and, although Washington also kept many satellite programs classified for years, their success, increasing needs and the growth of technology expanded the use of space in the military and, increasingly,

Space-based systems give nations an unsurpassed global reach. Space satellites can influence conditions for potential conflict. Denying access to American and other nations' space-based capabilities may be a goal for several countries. (USAF)

NASA's manned missions dominated early US space programs. Most people identified with America's efforts to place a man on the moon, but few knew about its extensive military satellite programs. Space Shuttle crews flew a few military missions, but the military has not developed a dedicated manned spacecraft. (NASA)

the civilian worlds. Military space systems still conduct the critical mission of protecting the nation from attack, but they have also stretched into the realm of supporting military operations by air, land and sea, particularly from the 1991 Persian Gulf War onward. During that conflict satellites enabled weapons systems to allow precision attacks that greatly aided strategic bombardment and tactical operations. Communications, navigation, imagery, early warning and other functions also flourished and supported a rapid victory over Iraqi forces.

The United States has developed a growing and sophisticated constellation of many military, commercial and civil satellites. Unlike with a terrestrial land fortress, countries that employ these satellite constellations do not have to worry about national boundaries. Similarly, spacecraft can stay aloft for years, unlike aircraft or naval fleets. Technology has allowed spacecraft to remain largely unmanned. Military space systems, as a result, have ground controllers stationed around the globe conducting actions on satellites. In the past, military satellite activities were cloaked in secrecy. Today, their use is not only common, but a necessity to conduct military operations.

Protecting existing space systems has become a hotly debated issue. Regional foes and competitors view America's space systems as a target that, if disabled, could limit Washington's ability to conduct global military operations. Ensuring an adversary cannot destroy or deny the nation's military and civilian access to satellites or space is an expanding requirement for the Pentagon. Nations like China, Russia or Iran may seek ways to counter space-based capabilities. These efforts include actions from destroying orbiting spacecraft to jamming their signals. US military space forces must design systems to protect these assets. Navies ensured pirates did not harass commercial maritime activities; forts defended

Space assets affect everyday life. Despite their military origins, satellites provide commercial and civilian uses from communications to weather reporting. This Defense Meteorological Satellite Program spacecraft provides continual weather reporting. (USAF)

Fears of a surprise nuclear strike on the US encouraged military space development. A nuclear attack could devastate large areas of the country for the first time. Detection of an enemy's capability to conduct an attack was vital. (National Archives)

settlements and economic development; and air forces secured the nation against a nuclear bomber threat. Military leadership now faces the challenge to do the same for space forces.

The role of military space operations and missions has evolved from a supporting function to one where weapons might actually be deployed from orbit. These capabilities may include manned space operations, a function that the Air Force does not conduct today, whilst the expansion of dropping precision-guided munitions could have significant impact on how America conducts future military operations. The speed and range of conducting operations from space could give national and military leaders a wide range of options.

Today, other nations are developing their own military space capabilities. Nations that already rely on communications, navigation, imagery, warning and other activities are working on ways to protect these systems and expand their constellation, which may potentially lead to conflict. Damaging a rival's spacecraft may have serious side effects. If an adversary decides that the cost of a preemptive attack outweighs the benefit of space-based systems, then it might unleash a massive attack in space. This might result in a tremendous reduction of space system capability to innocent parties. Loss of these systems could literally change the face of warfare in the future.

Chronology

June 6, 1942	Germany conducts the first successful launch of a V-2 ballistic missile. This demonstrates a potential booster system to explore space.
April 16, 1946	The US launches a German V-2 from White Sands Missile Range, New Mexico.
May 2, 1946	RAND study, for the Air Force, which demonstrates the possibilities of developing a man-made satellite.
August 29, 1949	Soviet Union explodes its first atomic bomb.
August 12, 1953	Moscow detonates its first thermonuclear device.
July 1, 1954	Air Force officials create the Western Development Division.
November 27, 1954	Washington approves the building of a reconnaissance satellite under WS-117L.
October 4, 1957	Sputnik I orbits the earth.
February 28, 1959	Air Force engineers launch Discoverer I.
April 1, 1960	NASA launches first weather satellite, Tiros I. This civilian satellite would later inspire the Department of Defense to produce the Defense Meteorological Satellite Program.
May 24, 1960	Air Force crews put MIDAS early warning satellite into orbit.
August 18, 1960	Vandenberg crews launch first successful CORONA mission, Discoverer XIV.
May 28, 1964	Program 437 ASAT made operational.
June 16, 1966	The Defense Satellite Communications System program first pushed into orbit.
April 25, 1967	US Senate ratifies Outer Space Treaty.
November 6, 1970	First Defense Support Program satellite employed.
September 1, 1982	Air Force Space Command created.
March 23, 1983	Strategic Defense Initiative ("Star Wars") announced.
January 28, 1986	Space Shuttle *Challenger* destroyed in flight.
January 17, 1991	Space systems used extensively in Operation *Desert Storm* aerial combat missions.
March 9, 1994	Full constellation of 24 Global Positioning System satellites is established.
November 20, 2002	First Delta IV EELV launched.

Liftoff into the deep black: military space design and development

America's military space program developed in response to a perceived threat from the Soviet Union after World War II. The US had seen profound changes in military technology through the 1940s: ballistic missiles, radar, the atomic bomb and jet propulsion had all been introduced. Protecting America by relying on oceanic barriers, coastal defense artillery and the Navy was problematic at best. America survived Pearl Harbor. Could it afford to gamble with a future surprise nuclear attack? Clearly, the December 7, 1941, Pearl Harbor attack illustrated a lack of strategic intelligence and warning capability that could have mitigated the raid.

Unfortunately, the end of World War II did not eliminate all threats facing Washington. The Soviet Union had demonstrated its will and ability to expand communism worldwide by taking control of Eastern Europe and supporting revolutionary movements. Although the US held a nuclear monopoly and maintained an ability to deliver it with bombers, the Kremlin was seeking ways to challenge American and Western European national interests. This effort included the development and production of nuclear weapons and delivery systems. Moscow demonstrated its ability to challenge the Free World by blockading Berlin in 1948, exploding an atomic bomb in 1949, supporting a war in Korea from 1950, developing strategic bombers and espousing political rhetoric that included global expansion. On August 12, 1953, the Soviets successfully detonated a thermonuclear device and later that year revealed their MYA-4 Bison bomber, their answer to the Air Force's strategic bomber fleet. America's sense of nuclear security had evaporated. A possible attack on the United States caused military

Ballistic missiles protected the nation from the growing Soviet threat and supplied Washington with a ready source of space launch boosters. Atlas was America's first intercontinental ballistic missile; it put an astronaut into Earth orbit under the Mercury program and pushed many spacecraft into space. This Atlas D carries a MIDAS early warning satellite into orbit. (USAF)

The Atlas intercontinental ballistic missile served as an early workhorse to lift the burgeoning military space program. Although developed in the 1950s, Atlas still serves today and will do so well into the 21st century. (USAF)

and political leaders to search for new ways of finding out what the Soviets would do next. The Joint Chiefs of Staff could not guarantee that Soviet bombers, or worse, ballistic missiles, would not unleash a "bolt out of the blue." Intelligence gathering through aerial and human approaches was difficult with regards to the Soviet Union due to both the closed nature of its society and its vast territory.

The United States Air Force (USAF) and Navy did not possess any aircraft that had the range or ability to fly at high enough altitudes to avoid Soviet air defenses. Although Washington used high-altitude balloons with cameras, this method was highly inaccurate and inconsistent. Air Force and Central Intelligence Agency (CIA) officials would eventually build the Lockheed U-2 aircraft; however, a more secure way of gathering information was to use reconnaissance and surveillance satellites.

By 1946, the RAND Corporation, a private "think tank" created by the Air Force to examine scientific and technical issues, was commissioned to conduct feasibility studies into creating Earth-orbiting satellites. RAND recommended that the United States pursue orbiting satellites to support the development of long-range rockets that could be used as ballistic missiles, advance military capability, aid scientific research, and help the nation psychologically and politically. Satellites could support two key areas: surveillance and weather.

On November 27, 1954, the USAF's Air Research and Development Command (ARDC) issued System Requirement 5 for an Advanced Reconnaissance (Satellite) System. By March 16, 1955, Headquarters Air Force would approve the requirement and start development of Weapons System (WS)-117L, the

The world's first orbiting satellite: Sputnik

Throughout the early to mid-1950s, the US scientific and engineering community believed that it had a tremendous advantage in satellite design and space launch capability. Although ballistic missile development experienced agonizing problems, no other country seemed to rival the US. On October 4, 1957, a national crisis developed when the Soviet Union launched Sputnik I. Soviet engineers put the 83kg satellite into orbit aboard an R-7 Soviet ballistic missile from the Baikonur space complex – the event shocked the world. Sputnik I transmitted signals around the globe for three weeks. On November 3 the Soviets surprised American space experts further by launching Sputnik II, which contained the world's first living organism to orbit the Earth – a dog called Laika. America answered the satellite challenge on January 31, 1958, with the launch of Explorer I.

The US failure to launch its Vanguard space satellite before Sputnik led to allegations of mismanagement in the ballistic missile and space programs. Although Sputnik shocked the country, the United States responded by increasing funding and putting more emphasis on ballistic missile and space efforts. This allowed the Air Force to initiate several new projects that would pay handsome dividends in the future.

RIGHT The Titan II ballistic missile carried the largest nuclear yield for the USAF. This ballistic missile served as the basis of a family of SLVs and, after the USAF removed them from service, they operated as space boosters. Launch crews tested this Titan II at Vandenberg AFB. (USAF)

BELOW Early space and missile development was experimental and led to spectacular launch failures. Early test launches, such as this Titan ballistic missile, were critical towards building an effective space program. Air Force crews used obsolete Titan ballistic missiles as boosters and their derivatives still serve today to put satellites into orbit. Unfortunately, this Titan test blew up on the launch pad. (USAF)

US's first attempt to create a reconnaissance satellite. It was envisioned that WS-117L would provide a means to conduct continuous surveillance of selected Soviet sites by attaining an orbit, receiving and executing commands from the ground, and transmitting information to a ground station. The Strategic Air Command (SAC), responsible for the Air Force's nuclear bomber fleet and eventual control of land-based ballistic missiles, viewed surveillance and reconnaissance as a vital instrument to gauge the Soviet Union's strategic intent. WS-117L was supposed to gather information on airfields and ballistic missile launch sites, conduct electronic intelligence gathering and report on weather conditions. ARDC would build WS-117L under its Western Development Division (WDD), whose engineers were also responsible for developing and producing ballistic missiles, a link that would support space booster employment and related areas.

Militarizing space

The United States began space system development with a particular goal: protect the nation from a Soviet attack. Reconnaissance and surveillance were the priorities and these systems radically altered CIA and other intelligence agencies' opinions on Soviet ballistic missile, bomber, naval and nuclear developments, as well as operational activities. Before these satellites, most intelligence analysts believed the Soviets would

LEFT The military space program evolved from simple cameras in orbit to sophisticated ballistic early warning satellites. The infrared sensor on this DSP satellite has allowed the nation to extend its warning system globally. (USAF)

need years to produce a working intercontinental ballistic missile (ICBM) that could harm America. Besides, American ICBM efforts had floundered through the early and mid-1950s and the country had greater technical capabilities than Russia; scientists found it inconceivable that Moscow would build a long-range ballistic missile or place an item in space orbit before America. On October 4, 1957, Moscow shocked the world with its launch and successful orbiting of Sputnik I. This demonstration of a simple satellite pushed the US into expanding both its ballistic missile programs and WS-117L. If Moscow could send a payload into orbit, then it could launch a nuclear-armed ballistic missile. Additionally, the Soviet Union could also create an orbital bombardment system to rain nuclear terror from above. The shock of seeing a Soviet satellite system in space also highlighted the inability of American strategic intelligence to perceive this threat. The Eisenhower administration needed better intelligence, warning, communications and a host of new systems to counter this new development.

Government engineers and scientists proposed a series of both manned and unmanned systems to not only provide reconnaissance, surveillance, and warning, but also conduct orbital bombardment and anti-satellite (ASAT) operations. However, the first priority was WS-117L and its photoreconnaissance version became the CORONA program. Launched from a Thor-Agena booster, CORONA provided a comprehensive survey of the Soviet Union, People's Republic of China (PRC) and other areas. Thor was an intermediate range ballistic missile (IRBM) developed by WDD. On August 18, 1960, Discoverer XIV (Discoverer was the codename for the CORONA program) carried the first fully successful satellite to return film from space. Air Force crews from Hickam Air Force Base (AFB), Hawaii, flying a modified C-119, captured a returning film capsule in mid-air. The CORONA program included 145 launches and photographed the Soviet Union until May 31, 1972. The program allowed Washington to reduce U-2 missions, especially after the Soviets downed one in 1960, and provided a wealth of information that dispelled myths about Soviet ballistic missile superiority, the so-called "missile gap." This program demonstrated the feasibility and value of military space programs.

Founder of the American military space program: Bernard A. Schriever

One could consider General Bernard A. Schriever as the founder of America's military space program. Schriever was a career Air Force officer who was born in Bremen, Germany, on September 14, 1910. He graduated from Texas A&M University where he received an engineering degree and was commissioned into the Army. Schriever transferred to the fledging Air Corps, but he concentrated on technical and engineering assignments. In World War II he was a B-17 pilot and served in several technical and support assignments.

After the war, Schriever held assignments that involved the development of several new weapons. He headed the Atlas ICBM program, activated the Western Development Division, took responsibility for the WS-117L project and commanded the ARDC and its successor, Air Force Systems Command. ARDC was responsible for developing air, missile, space and other major Air Force programs. Schriever strengthened the Air Force's development, research and acquisition activities that contributed significantly in advancing new technologies for space and other programs. By April 1961, he was a four-star general and oversaw the military space and ballistic missile programs. General Schriever advocated the use of ballistic missiles and space systems during this critical stage of development. This advocacy was instrumental in gaining program approval and funding. Schriever retired in 1966; however, he continued service to the nation as a consultant on a host of committees concerning intelligence, science, technology and ICBMs. He died on June 20, 2005.

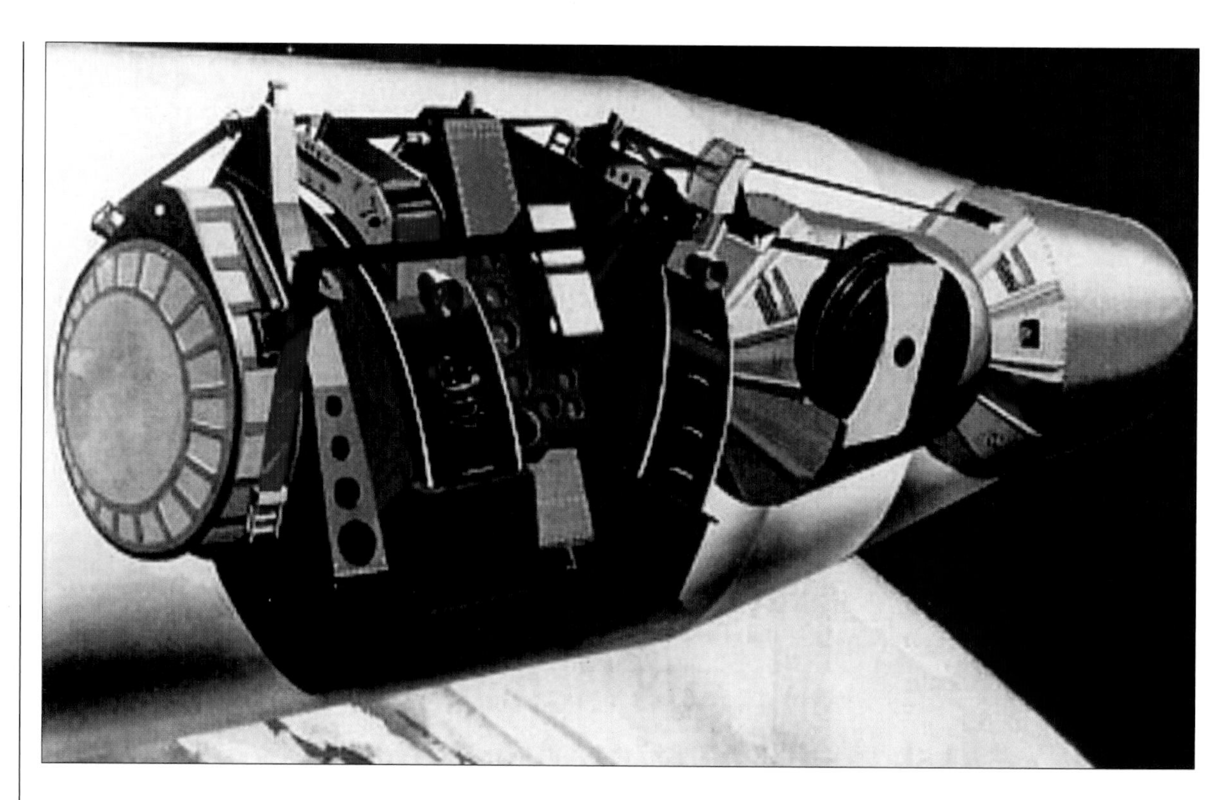

The secretive CORONA program was the first US photo-reconnaissance satellite. It gave a safer and more comprehensive coverage of the Soviet Union than manned aerial systems like the U-2. After its imagery mission, ground controllers would order the film capsule ejected over the Pacific Ocean for recovery. (NRO)

Other WS-117L systems followed CORONA, but were not as successful. Engineers designed the Satellite and Missile Observation Systems (SAMOS), another photoreconnaissance satellite, to take images and electromagnetic data, but the program had limited success due to technical problems in its electro-optical scanning system. An Atlas ICBM booster launched SAMOS. Another WS-117L offshoot, the Missile Defense Alarm System (MIDAS), was designed as a constellation of satellites to detect ICBM launches by their infrared (IR) signatures from the missile's exhaust. This early warning satellite network should have provided a near-instantaneous detection mechanism for Washington to initiate a retaliatory nuclear strike. However, MIDAS' original constellation of eight polar orbiting satellites was never completed due to technical failures. In 1962 the MIDAS program was continued as a research program to explore new technologies the in hope of one day developing into a working system. National leaders saw the value of building not only a MIDAS-style program, but also a series of ground radars called the Ballistic Missile Early Warning Systems (BMEWS). MIDAS' successor and BMEWS would provide their part of America's ability to prevent a "nuclear Pearl Harbor."

Demands for providing up-to-date information about the Soviet Union expanded. Military planners required accurate weather reporting to conduct a host of activities from training missions to preparing a nuclear strike on Moscow. Similarly, advances in ballistic missiles and nuclear-powered submarines allowed the Navy to launch a nuclear attack from the seas. However, this new capability required precise navigation. Satellite programs, such as Transit, helped Polaris submarines adjust their missiles' guidance systems. The Army supported development of Advent communications satellites that linked forces worldwide and reduced the need for radios and telephones. The Air Force explored satellites that detected nuclear detonations under the Vela program. This program gave analysts valuable data on a nation's scientific advancements that might influence arms control and verification. These force support measures provided military authorities with the ability to improve their targeting, better integrate their forces, coordinate activities and plan actions.

In the early 1960s, satellites demonstrated their true value. During this period, the Secretary of Defense Robert S. McNamara centralized military space systems under the Air Force in March 1961. Now, ballistic missiles, except submarine-launched ballistic missiles (SLBMs), and space systems fell under the one service. Ballistic missile development supported space launch activities, while space systems provided key information for military operations, especially nuclear ones.

Technology, funding, and political concerns had applied the brakes to some military space systems. Space launch boosters were still unreliable and frequently blew up on launch pads. Limited payload weights and component reliability for satellites constrained their capabilities and longevity. Ballistic missile and strategic nuclear programs had taken center stage for funding, space systems were still secondary. As the Vietnam War heated up, any excess funding was pushed into producing conventional forces. However, the most important limits were due to political concerns. The nation wanted to continue using space for reconnaissance and surveillance, communications and other missions, but Washington feared expanding an arms race into space. Washington could ill afford Soviet nuclear orbiting bombardment systems over the country. Negotiations with Moscow might eliminate this threat yet still allow for the exploitation of other military space systems. At the same time the National Aeronautics and Space Administration (NASA) had started to gain control of many of the manned and unmanned space programs. Civilian space programs began to dominate the field, at least in the public's eye. The success of CORONA and other intelligence gathering satellites also led a successful attempt by the CIA, under the guise of the National Reconnaissance Office (NRO), to wrest control of these assets from the USAF. Military and key national space systems started to become highly classified "black" programs due to their sensitive nature. These systems became the future eyes and ears for not only military, but also for key international programs supporting treaty negotiations.

Sending military space systems into orbit

Developing operational space systems created difficult challenges for scientists and engineers. The difficulties of propulsion, environmental hazards and numerous other problems forced the USAF and others to develop a new science to solve them. Space has no shape or substance and it allows almost limitless maneuver and visibility; however, systems must operate in a vacuum and engineers had to design satellites and components to withstand swings in temperature, high levels of radiation and micrometeoroids that can damage sensitive instruments.

Strategic reconnaissance was a vital space mission advanced under the Eisenhower administration. President Dwight D. Eisenhower revealed this Discoverer XIII capsule recovered in the Pacific Ocean at a press conference on August 15, 1960. (USAF)

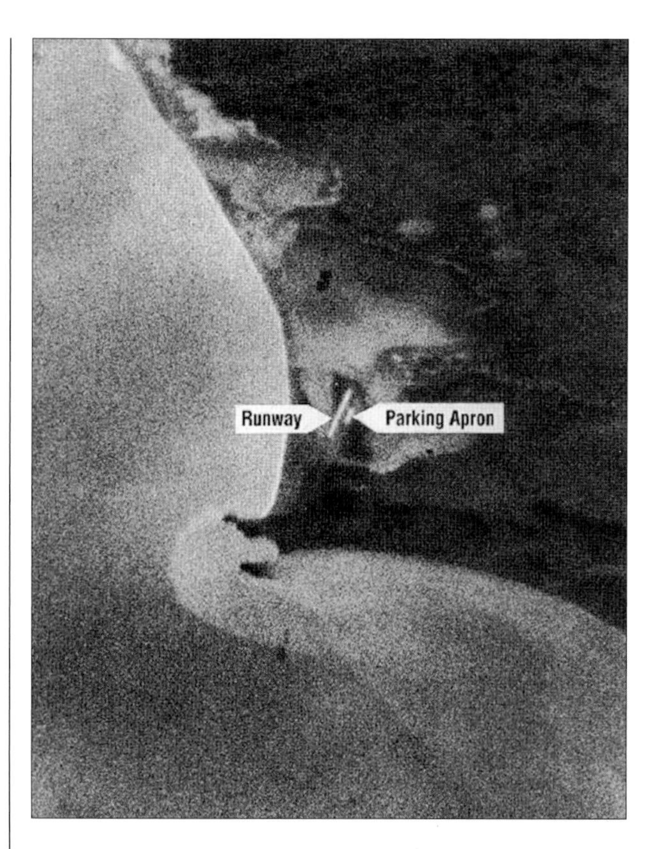

ABOVE CORONA imagery enlightened Eisenhower and his successors with prime intelligence data. Soviet airfields like this one at Mys Shmidta were used to prove there was a lack of a long-range bomber threat to the United States. The feared "bomber gap" between the US and Soviet Union was dispelled. Some national and military leaders thought the Soviet had built and deployed more strategic bombers than America, but CORONA helped disprove this idea. Intelligence analysts received this image made on CORONA's maiden mission. (NRO)

Space begins at an altitude of about 100km above the Earth's surface. International law recognizes space to begin at the lowest altitude where an orbit around the Earth is sustained, the region where vehicles do not have to rely on aerodynamic forces, and the principles of atmospheric lift and drag become irrelevant. Satellite designers have to use great care to ensure proper orbital mechanics are addressed. Depending on a satellite's mission and planned duration, orbits can take several forms and different orbits affect a satellite's area coverage, ability to communicate with a ground station, and length of time in space.

Orbits normally fall under several categories: circular, elliptical and by altitude and inclination. A satellite must maintain sufficient velocity equal to the force of the Earth's gravity to maintain an orbit. If the velocity is greater than the pull of gravity, then the vehicle will escape orbit, and if it is lower the vehicle will become suborbital and force reentry back to the surface. Circular orbits allow a satellite to make one complete revolution around the Earth every 90 minutes at an altitude of *c.*200km at around 8km per second. If military commanders want a continual global perspective, then a satellite constellation in circular orbits could provide comprehensive coverage. Elliptical orbits have a perigee (orbital point closet to Earth) and apogee (a distance farthest from the surface). Engineers use these orbits to cover a particular area of Earth, getting close to a region and then moving further out.

One can also classify orbits by altitude or inclination. Satellites can be placed into specific altitudes: low Earth orbit (LEO, 100–500km), medium Earth orbit (MEO, 500–36,000km), geosynchronous Earth orbit (GEO at 36,000km), and

RIGHT Although primitive by today's standards, early satellite imagery, like this early CORONA snapshot of the Pentagon taken on September 25, 1967, supplied valuable information to the analysts. (NRO)

Specially equipped C-119 aircraft recovered CORONA film capsules. Crews from Hickam AFB, Hawaii, conducted these missions. If not captured by the C-119, the capsules were designed to sink in the ocean to avoid them falling into Soviet or Chinese hands. (NRO)

high Earth orbit (HEO, above GEO). LEO orbits are vulnerable to atmospheric drag, but are relatively close enough to the surface to photograph images or to communicate with ground stations. Getting satellites into LEO is easier to accomplish and, consequently, the greatest density of vehicles is found here. Photoreconnaissance, electronic surveillance, navigation, weather and communications satellites populate LEO. MEO satellites are very stable, but face the Van Allen Radiation Belts that can degrade systems over time. MEO is popular for navigation and some nuclear-detection satellites. GEO orbits correspond with the time it takes the satellite to rotate around its axis; this creates the perception that the satellite maintains a constant position over a particular area. These orbits create an excellent path for surveillance and reconnaissance satellites, early warning and some communications spacecraft. A maneuver to put a satellite into a GEO is the geosynchronous transfer orbit (GTO). A booster puts a payload into LEO and is placed in a "parking orbit" until it is in a proper position. A booster rocket then sends or transfers the payload to a GEO. HEO paths have more maneuver room, but are very distant, which gives some concerns about image quality and signal reception.

Orbits are also classified based on their inclinations. Launch crews can send a satellite into an equatorial orbit, sun synchronous or polar flight path. Equatorial orbits offer wide coverage of areas since their line of sight offers good north to south contact. Sun synchronous satellites pass over certain areas, at prescribed times, every day. Some reconnaissance and communications satellites can use these orbits if only periodic coverage is required. Polar satellites allow satellite line-of-site coverage of appropriate hemispheric areas.

Before a satellite can achieve orbit it must escape Earth's gravity. Creating sufficient propulsion to lift a payload into space and velocity to put the satellite into orbit requires support, and unmanned systems require launch and ground support to function. Most military launch vehicles are derived from ballistic missiles—ICBMs or IRBMs. In the past, the main military space launch vehicles (SLVs) were the Atlas, Titan and Thor. The first successful military space launch occurred on October 11, 1958, with a Thor carrying a lunar probe. An Atlas developmental missile that carried a communications package went into orbit by December 18, 1958. As military space systems became more complex and required greater weight, the manufacturers had to create larger SLVs. Titan expanded into a system with two solid rocket booster strap-on motors. Atlas, first launched in the late 1950s, continues as an SLV to this day: military space systems are still launched by expendable SLVs only. The USAF had planned on using the Space Shuttle for most of its military missions, but only a few military satellites were deployed by this system. After the Space Shuttle *Challenger* accident on January 28, 1986, Air Force officials assigned military satellite launches to expendable SLVs indefinitely.

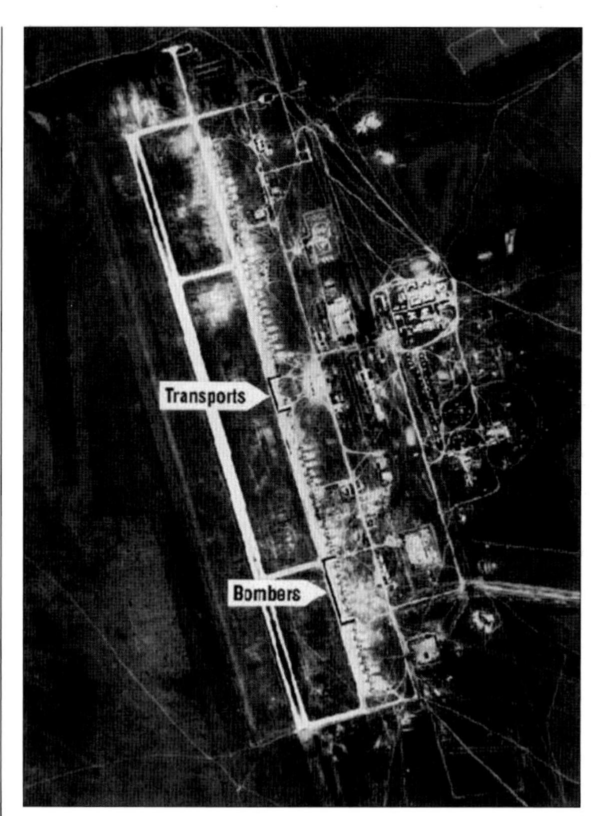

As technology improved, national leaders relied on satellite imagery for many key decisions. This CORONA picture from 1966 shows a vast improvement over the first set of photographs from 1960 and analysts could determine the type and number of aircraft on this Soviet airfield. (NRO)

Air Force space elements

Initially, the ARDC controlled space activities succeeded by the Air Force Systems Command. The Pentagon's growing awareness of the importance of space operations was confirmed by the creation of the Air Force Space Command (AFSPC) in 1982. Based at Peterson AFB near Colorado Springs, Colorado, this command controls not only USAF satellites, but its ICBM force as well. Once dominated by ballistic missile development, space has now become the lead element of the relationship. AFSPC operates military satellites; overseas ground-based missile early warning radars; national launch centers and ranges; space surveillance; and mans ballistic missile sites. AFSPC falls under the control of the US Strategic Command, which is responsible for global strike and ballistic missile defense, as well as controlling space and information operations.

There are two main military launch centers: Vandenberg AFB, California, and Cape Canaveral Air Force Station (AFS), Florida. Both launch sites have tested ballistic missiles and participate in military and civilian space launches. As these launches were highly experimental at the outset it was decided that they had best take place away from populated areas. Vandenberg AFB and Cape Canaveral AFS allowed launch crews to launch their SLVs or test missiles over the ocean, thus avoiding any risk to the civilian population. Vandenberg launch crews sent satellites into a polar orbit instrumental in observing the Soviet Union. Vandenberg is also the sole test bed for USAF ballistic missiles and has six times the acreage of Cape Canaveral. Cape Canaveral sends military payloads into equatorial and geosynchronous orbits and is adjacent to Kennedy Space Center. Unfortunately, both launch sites are subject to environmental damage: earthquakes are common in California and Florida is subject to hurricane and storms. Vandenberg was assigned the mission of serving as a second Space Shuttle launch and landing site, but it returned to expendable rockets after the *Challenger* explosion. In the past these launch sites were susceptible to intelligence gathering by ships loitering off shore: today they are also both closely monitored by foreign satellites.

A key ingredient of any military space operation is the satellite control system, which includes ground receiving and transmitting antennas, data processing, communications and operations centers. AFSPC operates a series of tracking stations and control centers; most tracking stations are located in the continental United States, but tracking stations in Hawaii and Guam extend the reach of military satellite control through the Pacific. There are also stations in Thule AFB (Greenland), Oakhanger (England) and Diego Garcia in the Indian Ocean. Air Force personnel operate satellite control facilities located at Schriever AFB close to Colorado Springs and at Onizuka AFB near Sunnyvale in California (which is slated for closure). Schriever AFB acts as the Consolidated Space Operations Center for all satellite control while another facility at Buckley AFB near Denver, Colorado, acts as communications processing center for a host of early warning and communications functions.

There are many other organizations that actively plan and operate space systems. The Naval Network and Space Operations Command in Dahlgren, Virginia, controls Navy space communications systems. The Army's Space and Missile Defense Command oversees Army-specific communications and space activities, as well as ballistic missile defense activities.

Other key US space organizations include several intelligence agencies, foremost among them the CIA, whose Directorate of Science and Technology designs and builds systems to support requirements for a vast array of activities. The National Geospatial-Intelligence Agency (NGA) uses space imagery and other information to produce maps and other products to support military forces. The NRO designs, builds, and operates reconnaissance satellites. National Security Agency (NSA) personnel are charged with protecting communications and gathering foreign intelligence communications. Although the CIA and NRO are not solely used for military purposes, USAF crews launch them.

ABOVE Frequently, additional booster systems push spacecraft from a "parking orbit" to a higher one. Space Shuttle crews use this Inertial Upper Stage to push satellites from LEO to different orbits. (DOD)

How satellites work

Aside from specific equipment to support its particular mission, each satellite shares certain common structures and functions. These subsystems provide a support structure, communications, propulsion, attitude control, power systems, thermal control and a data system. Satellite designers must ensure that each system can operate independently for years and limited access after the satellite is in orbit forces designers to add redundant hardware and software.

A space system structure protects its components during construction, launch and operation in orbit. Ground station personnel must have a communications link to transmit and receive data from the orbiting satellite to execute commands or download information. The satellite must also have an ability to modify its

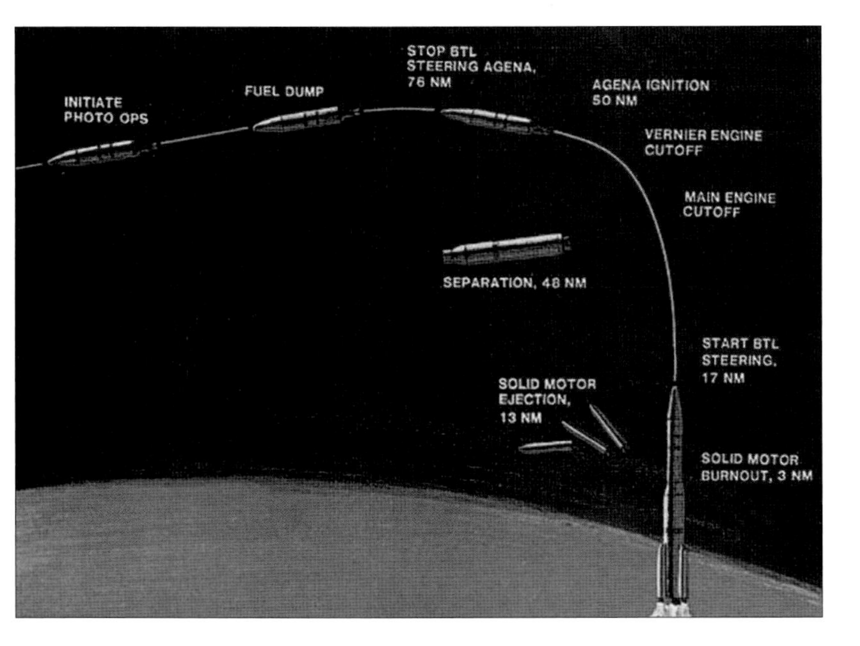

LEFT This sequence of activities illustrates the major steps taken to put a CORONA spacecraft into orbit and start its mission. (NRO)

RIGHT **Detection of ballistic missile launches**
The US relies on a series of satellite- and surface-based radar systems to detect ballistic missile launches. National and military leadership receive early warning from the DSP satellites in orbit. This DSP spacecraft has detected North Korean Taepo-Dong ballistic missiles fired east towards the United States. A missile interceptor, armed with a kinetic kill vehicle, from Vandenberg AFB in California attempts to defeat this Noth Korean threat.

velocity to change orbits or to maneuver, and propulsion systems can carry either a solid- or liquid-fueled rocket system to conduct these movements. Attitude control capabilities allow ground controllers to point the satellite in the correct position to conduct its mission. All these components require electrical power to operate and each satellite needs to either produce electrical power on its own or use stored power. Frequently, satellites use banks of photovoltaic cells or solar arrays that convert light to electrical power. Thermal control systems protect components from external heating and cooling due to the space environment; they also control internal heating allowing the various subsystems to operate. The last major satellite components include data systems to track the satellite, convert signals, conduct command and control functions, perform onboard computing, and allow any autonomous operations.

Satellite design and testing ensures components produce a maximum capability that is reliable at the least weight. Reduced weight allows satellite designers to increase system capabilities, as a heavier payload requires a larger SLV or placement in an orbit that requires additional propulsion, thus reducing other component capabilities.

Satellites also do not exist in a benign environment. Despite being in a vacuum, space has several hazards ranging from electromagnetic radiation that affects satellite temperature to meteoroids, solar wind, cosmic rays and solar flares. These conditions create problems for space operations including control and tracking issues due to interference, damage to satellite components because of excessive or stray voltage, overheating, slow degradation of optical and other systems, satellite disorientation and communications anomalies.

These problems have led to the development of hardened systems to combat the environmental conditions. Additionally, the Air Force has had to create redundant systems to ensure emergency capability in case one or more systems does not function The satellite also needs to have the capability to repair and change operations by itself or with ground commands.

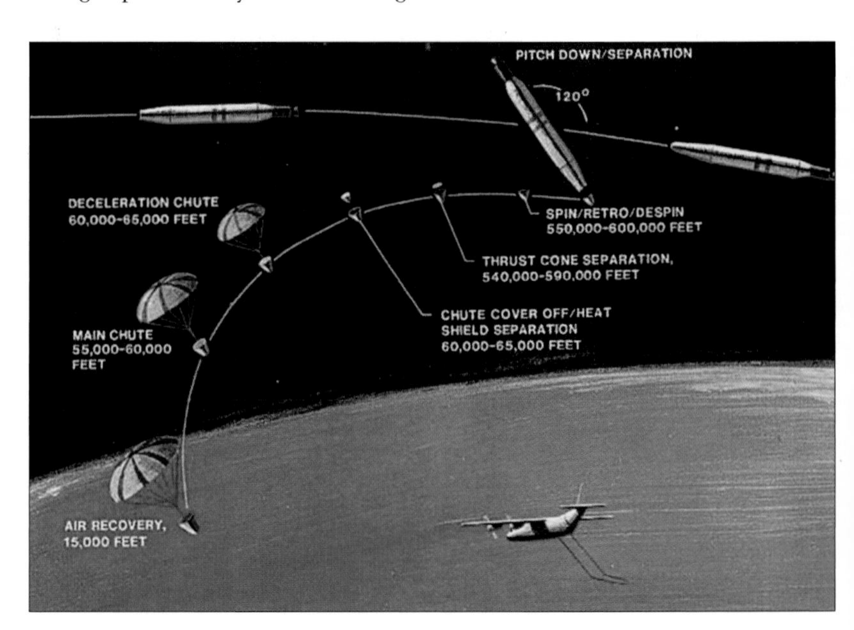

After its mission was completed or ordered to recovery, the film canisters would return to Earth. An Air Force C-119 (later C-130) would capture the reentry capsule. (NRO)

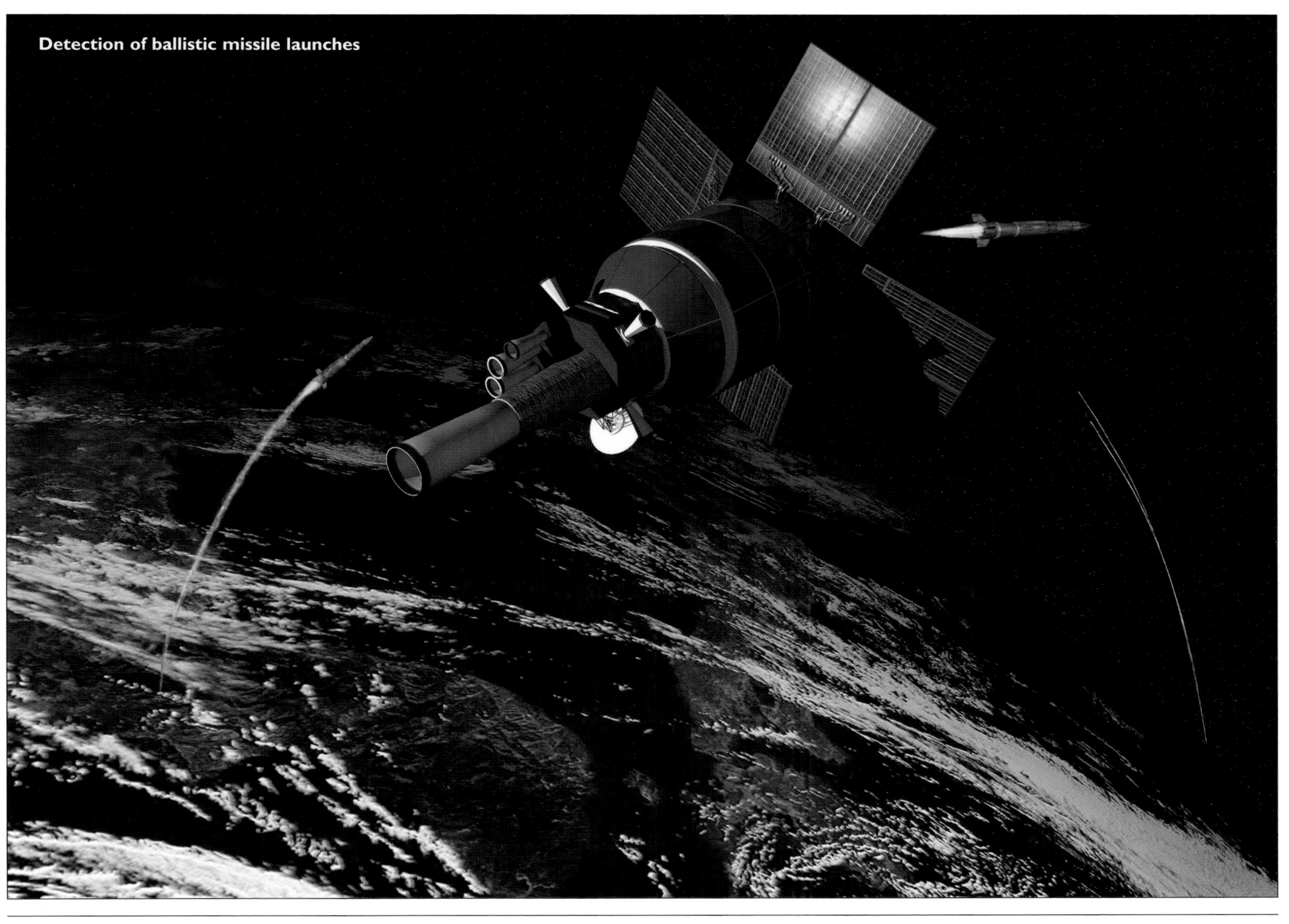
Detection of ballistic missile launches

Satellite design must balance requirements of function versus weight. Putting excessive or marginal systems onboard the satellite will cost a satellite operator extra expense and weight given strict SLV requirements. Satellites, like this GPS craft, must face extensive examination and testing. (USAF)

Satellites do operate in an independent manner while in orbit, but rely on a host of terrestrial supports. Although designed to accomplish a specific mission, frequently over a specific region, a ground control element can change the satellite's operating functions. They can shutdown satellite activities, modify operations, maneuver, repair a failed function by command and reprogram a computer system. Ground controllers receive and analyze satellite tracking data or telemetry. Using this information, these controllers can order appropriate changes to operations.

Major US space functions

Early military space efforts focused clearly on the perceived Soviet threat from a nuclear ballistic missile or bomber strike. While reconnaissance, surveillance, and early warning functions took center stage, the military conducted significant space research on manned and unmanned activities. Manned experiments with hypersonic vehicles and other potential subsystems to put military personal into space continued from the 1950s through the 1960s. In cooperation with NASA, the Air Force and other services performed valuable test programs and the X-15 was the first vehicle to reach the boundaries of space with a maximum speed of 7,300kmph in 1967. Several test pilots earned astronaut wings and NASA's first set of seven astronauts for the Mercury orbital program were all military officers; the military seemed poised to expand from unmanned to manned space systems.

Debate within Washington circles raged; should the military expand into a manned space program? This would be a natural extension from aerial flying to space. However, NASA was already building a series of programs that would eventually land a man on the Moon by the end of the 1960s and duplication of two separate manned space systems appeared excessive. The nation was already conducting a massive strategic nuclear build-up program and technical concerns about manned space flight continued. However, the USAF now faced the rising cost of conducting Southeast Asian combat operations whilst questions about a potential space arms race also clouded the issue of extending the military's reach into manned space. Unmanned systems seemed more benign and less expensive. Military cooperation with NASA would continue, but military manned space flight stalled until the advent of the Space Shuttle. The early American military space effort, albeit decentralized and disorganized, offered the potential ability to dramatically improve targeting capabilities, early warning, communications and other military support activities. The program crashed abruptly with the *Challenger* explosion. Today, the possibility of military manned space flight has remerged and, while military space operations focus currently on unmanned systems, future systems may create an opportunity for a possible return to military manned space missions.

After design, development and assembly, satellite engineers put spacecraft through exhaustive tests. This early DSCS satellite illustrates the exacting signals tests in an anechoic chamber. (USAF)

Although military space functions are consistently evolving in the face of a changing geopolitical and technological environment, some missions have remained consistent over time: **reconnaissance and surveillance** provides much-needed information for everything from targeting to arms control. Space systems can observe activities on the surface, at sea, in the air or orbiting in space. Near-instantaneous information can support ongoing operations or trace particular events over time to develop a response, like a country developing a new military capability. **Early warning** provides the government sufficient strategic warning in the case of a ballistic missile strike on the country and allows time for an adequate response. A growing mission in this area is theater ballistic missile warning. Combatant commanders must consider a growing threat of battlefield theater ballistic missiles from SCUD and derivative delivery vehicles on fielded forces. Satellite systems can provide information to allow analysts to calculate launch locations, time and likely impact points that can help missile defense and allow targeted areas to defend themselves. **Communications** allows commanders to transmit and receive information from fielded forces. Space communications networks give worldwide access from the president to an individual military member. **Navigation and timing** data gives forces three-dimensional position and data to calculate vehicle velocity. This ability provides invaluable guidance for aircraft or weapons delivery, land maneuver, communications network synchronization and other missions on or above the Earth's surface in all weather conditions. Space systems present commanders up-to-date information on crucial **weather** conditions. These satellites support a wide range of military planning from theater campaign to aircraft sorties. **Remote sensing and geodesy** satellites take measurements of the Earth's surface to allow analysts to construct topographic maps, ascertain geological conditions and assemble hydrographic information. These satellites use imagery and multi-spectral sensors to determine conditions from vegetation conditions to mountain heights. Military commanders can use infrared and multi-spectral analysis for surveillance against enemy ground operations.

Technological improvements, changes in political leadership, expanded threats and the growth and reliance on space have forced changes in the

The first-generation ballistic missile-warning MIDAS system used a limited infrared sensor in 1962. Advances in technology allowed the Air Force to improve the infrared sensor on the DSP spacecraft to provide better detection. DSP continues to serve today. (USAF)

thinking about the military use of space in recent years. This now includes an expanded role for some functions believed infeasible a few years ago, particularly those associated with the issue of "weaponizing" space. A system of **missile defense**, ballistic and cruise, originating from space is not a new idea. Currently, several satellites conduct functions for early warning of ballistic missile attacks. New missions may include target acquisition, tracking and destruction of missiles. **Space control** allows a nation to both protect its own space assets, and also deny an enemy the use of space. Space control relies on the ability to watch and monitor hostile space activities as well as using offensive actions to negate hostile system and defend or protect proprietary assets. These actions may include using ASAT weapons, but also feature communications jamming, disabling certain components and other non-destructive means. **Force application** is a more contentious function for military space operations. Military commanders may one day deliver weapons or conduct combat operations in space. In the late 1950s with Sputnik, the White House and the Pentagon feared a Soviet unmanned multiple orbiting bombardment system (MOBS). Today, technology has advanced to suggest the use of hypersonic vehicles in a number of roles, such as strategic bombardment, rapid mobility, repair or replacement of disabled space systems and intelligence missions.

United States military space capabilities also include a number of activities to sustain current systems in orbit. **Space lift** allows the military to launch satellites and manned missions, military and civilian, into orbit. Launch activities at Vandenberg and Cape Canaveral conduct the bulk of military launches, but they also serve an expanding market for civil and commercial space systems. **On-orbit support** is the last major function and includes maintaining global remote tracking stations and space operations centers. Military and civilian personnel use this and other systems, including on-orbit systems, to track, communicate, operate and manipulate data from satellites.

Evolving space use

Once thought of as an exotic military curiosity, space systems became a common aspect of military forces for a number of reasons. The relaxation of military space from the "black" world to a wider population demonstrated the usefulness of reconnaissance and surveillance programs. Technology enabled an explosion of commercial and military systems from cell phones to wireless Internet, navigation and other devices that demanded added capacity, which now out supplies military capabilities. Threats from theater ballistic missiles and demands for precision military fire support have added to calls for more capacity from military space systems. Additionally, many non-military uses of space activities have aided diplomatic, economic, scientific and other governmental services that have improved a host of activities that have an impact on military operations. Finally, reductions in military personnel and the nature of military conflict that has forced a lighter, less on-site supported force has caused commanders to use advanced technology, including space technology, as opposed to manpower.

Operating in the deep black: space operations

The United States maintains a fleet of exclusive military satellites, but supplements them with several commercial and civil satellite systems. Although the USAF operates the majority of military satellites, other military and government agencies control selected systems. From 1958 to 2005, the majority of the United States' "military" space payloads concentrated on reconnaissance satellites. Intelligence agencies now control these satellites, though military organizations have become large users of these systems. Communications satellites are the next most popular satellite system. To date the US has not used bombardment or other weapons in orbit, though it has tested ASAT systems, as have other nations. Over the years NASA concentrated on manned space missions and scientific exploration including interplanetary activities and this has received more press attention; however, unmanned satellite programs have led space efforts.

The American military space program has gone from dominating the civil agencies involved in space to taking a subsidiary role; however, military and other satellite systems normally use many of the same SLVs from Vandenberg or Cape Canaveral to send their payloads into orbit.

Military SLVs

The USAF operates a host of SLVs, many of which have the same bloodline as older ballistic missiles—indeed the US has used ballistic missiles removed from service due to obsolescence or arms control agreements. The Atlas and Titan series have been the mainstays of American medium and heavy space lift, though the USAF and contractors have built other SLVs from ballistic missile components that have developed into an independent system. For example, aerospace contractors developed the Delta medium lift system from Thor. The Air Force also has increasingly used a number of independently produced SLVs. These include the Athena and the air-to-space, solid-fuel Pegasus and Taurus programs. The military has also used the Space Shuttle.

After the *Challenger* Space Shuttle disaster in 1986, the Air Force increased its fleet of expendable SLVs to enlarge its launch capabilities instead of relying solely on the Space Shuttle. The reliance on older-designed, refurbished ICBMs to serve as SLVs created concerns about their reliability. After a series of Titan launch mishaps, the Air Force decided to build a family of lower-cost, greater-capability SLVs that became the Evolved Expendable Launch Vehicle (EELV) program. The program centered on two older systems, Atlas and Delta, which engineers would redesign. Eventually, the Air Force will replace earlier model Atlas, Delta and Titan systems with the EELV boosters.

One of the longest and most successful SLV programs is based around the Atlas, the nation's first ICBM. Atlas is a liquid-fueled system that can

Launch crews had to take great care to emplace a satellite on an SLV. The spacecraft had to survive launch and then take several steps to deploy into orbit. Mistakes by the ground-support crew could turn millions of dollars of satellite development into space junk. (USAF)

Atlas—nuclear weapons carrier and space booster

The first US ICBM, Atlas, provided a tremendous strategic boost to the nation's strategic arsenal. The missile provided a fast retaliatory strike weapon that was less vulnerable than manned bomber fleets that the Soviets could destroy on the ground. This weapon system also gave the country its first heavy lift space booster for unmanned satellites and allowed America to orbit unmanned satellites and the Mercury astronauts. Atlas became operational in September 1959. The initial Atlas D stage-and-a-half ballistic missile could send its 1.44-megaton nuclear yield to a target over 10,700km away in about 43 minutes. Later versions could deploy a 4-megaton yield. Advanced technology and questionable readiness forced Air Force officials to remove Atlas from an operational status in 1965. Fortunately, the Air Force had begun using Atlas as a space launch vehicle, a role it served in for many years. This workhorse has supported interplanetary exploratory missions and launched communications satellites, early navigational satellites and many other payloads. Deactivated Atlas ballistic missiles found new life in the expanding civil and military space programs. Engineers modified the basic Atlas by adding longer fuel tanks and using upper-stage propulsion systems like Agena or Centaur. The Atlas E/F model space launch vehicles could put a payload of 3,265kg into a low Earth orbit of 185km. Atlas serves today with newer and more powerful boosters. Despite its almost 50-year-old basic design, the US plans to use the Atlas V and other variants as a booster in the 21st century. Its low cost and high level of reliability after years of use and modification provides a versatile space lift capability for the United States.

launch payloads into LEO and GEO. The nation first used it to launch a pre-recorded Christmas greeting from President Dwight D. Eisenhower into space in 1958. By 1986, the USAF had run out of obsolete Atlas ballistic missiles and Lockheed-Martin started building a new family of Atlas SLVs. The Atlas program now consists of the Atlas II, III, and V vehicles. Atlas V can lift up to 8,600kg into orbit, while the other systems can carry about half that payload.

The Delta SLV family has served the nation since 1959 and has successfully placed communications satellites into LEO, polar, GEO and other orbits. The Air Force uses Delta II SLVs for the highly successful Global Positioning System (GPS) program. In common with other SLVs payloads change depending on the orbit type, so Delta II can put a 5,000kg satellite into LEO, but only a 900kg object into a GEO. The trade-off between payload size and orbit is not insurmountable. Engineers can add solid booster strap-on motors and add a second stage for more capability to compensate for these constraints.

Advanced Titan SLVs, like this Titan III, supplemented the first-stage ballistic missile's liquid-fuel propellant system with two solid rocket booster motors. These SLVs carried large payloads, like photoreconnaissance satellites, into higher orbits. The Titan III served the nation for years. (USAF)

The Titan family of SLVs also evolved from a first-generation ballistic missile system. The Titan I and II system served as America's heavyweight ICBMs. The Titan I became operational in 1962 and was retired in 1965. Titan II crews went on alert in 1963 and protected the US with the missile until 1987. The liquid-fueled Titan II had a nine-megaton nuclear yield, but also evolved into an SLV for the NASA-manned Gemini earth-orbiting missions. Later versions of Titan would carry heavier payloads including reconnaissance satellites; They were also designed to carry the proposed, but never deployed, USAF piloted hypersonic vehicles and a manned orbiting laboratory. The Titan IV was the most powerful SLV in the USAF's inventory. Assisted by two strap-on solid rocket boosters, it could place a 21,700kg satellite into LEO and a 5,800kg object into GEO.

The US uses other SLVs for military payloads including the commercially developed solid fuel Athena I and II boosters. Athena has deployed a lunar probe and launch, space imaging, and has even launched from Kodiak, Alaska. All ground launch controls are self-contained in a 12m van. A more interesting SLV is the Pegasus and its larger brother, Taurus. The Pegasus is a delta-winged vehicle that can be airlifted to an altitude of about

11,600m. A modified L-1011 TriStar acts as a mother ship and launches the Pegasus. Taurus SLVs work on a similar basis but use a larger first stage based on the USAF's Peacekeeper ICBM. The Pegasus is reminiscent of the aircraft-deployed X-15 program. The Pegasus's builder designed the Taurus to carry a payload of about 820 to 1,320kg into LEO. The Pegasus, Athena and Taurus demonstrate a capability to place small payloads into orbit with relatively little support. This capability could one day allow commanders to put limited capability and longevity satellites into orbit near a battlefield. These tactical satellites could plug a vital gap in communications or imagery due to a lack of coverage or ASAT actions.

A typical launch sequence to place a satellite into GEO using a GTO should serve as a good illustration of how SLVs operate. If the US wanted to place a ballistic missile early warning satellite to monitor North Korea, then engineers might suggest using a GEO to ensure continual coverage over the area. Depending on the weight of the payload, the USAF might select the Atlas II, III or V. Supposing the Air Force used an Atlas V to deploy a newly designed early warning satellite, careful coordination between satellite and SLV launch crews is needed to ensure adequate support to allow the satellite to get into orbit. Due to its location, engineers could put the satellite over North Korea using Vandenberg AFB. Once engineers place the satellite on the Atlas V, launch activities proceed with ignition of its first-stage liquid-fuel rocket engine. After the rocket moves to maximum thrust, the Atlas V begins to lift off the launch pad. Up to five solid rocket boosters also ignite and burn for about a minute. A pre-programmed signal jettisons the expended solid boosters in pairs every two seconds starting after 90 seconds. The main rocket engine burns for 100 seconds and then reduces thrust to 95 percent.

The Titan IVB lifted heavy payloads into space. Titan IVs, like this one from Vandenberg, would carry NRO payloads into orbit. The SLV's large fairing indicates a rather large payload. The USAF has replaced the Titan IV with the EELV. (USAF)

This artist's rendition of a DMSP satellite illustrates the fragile components of a spacecraft. It has a solar array to collect energy and solar blankets to protect it from temperature changes. DMSP has served the military and public for years. (USAF)

RIGHT **GPS constellation at work**

GPS navigational capability has revolutionized precision attacks. Instead of using many bombing raids to destroy targets by mass, GPS has reduced the need to use many bombers to a single strike. This illustration portrays how GPS can guide a Tomahawk cruise missile from its submarine launch, transit and strike on a radar dish. GPS has allowed weapons designers to conceive of new ways to use munitions.

Four minutes after launch, the main rocket engine shuts down and separates from the payload and a second stage. The second stage is a Centaur liquid-fuel propulsion unit; this propels itself and the payload into a parking orbit. The Centaur then ignites for a second time after eight minutes in orbit to put the payload into the proper orbit. Computer controls separate the early warning satellite from the Centaur. After testing and equipment and system checkout, the early warning satellite is ready for operations.

Military satellites

The two most well-known purposes for which military satellites are designed are navigation and communications.

Military navigation satellites have advanced from the early Transit systems to today's GPS. The USAF sent the first GPS satellite into space in 1978 and this constellation of satellites provides precise, continuous three-dimensional position and time information for any point on the Earth's surface to anyone with a receiver device. By using the GPS constellation, military and civilian users can replace traditional maps or celestial navigation. Soldiers need not fear getting lost; ships can accurately calculate courses; airmen can use these systems to deliver extremely accurate ordnance or target. Since US policy is to allow open use of GPS, other nations' militaries and non-state actors, such as terrorists, can also use these signals.

A user determines his location by using signals from a minimum of four GPS satellites. The user's receiver unit manipulates timing differences from the transmitting satellites' signal to its reception in the receiver unit. These timing calculations are the basis to determine range and location. If a plane or ship is in motion, then it can also use the timing differences over position to calculate

During the Cold War, weather information was key to a host of NATO requirements. This DMSP image demonstrates the wide coverage that military satellites can give to a military commander. Today, nations or individuals can purchase this same information for a price. (USAF)

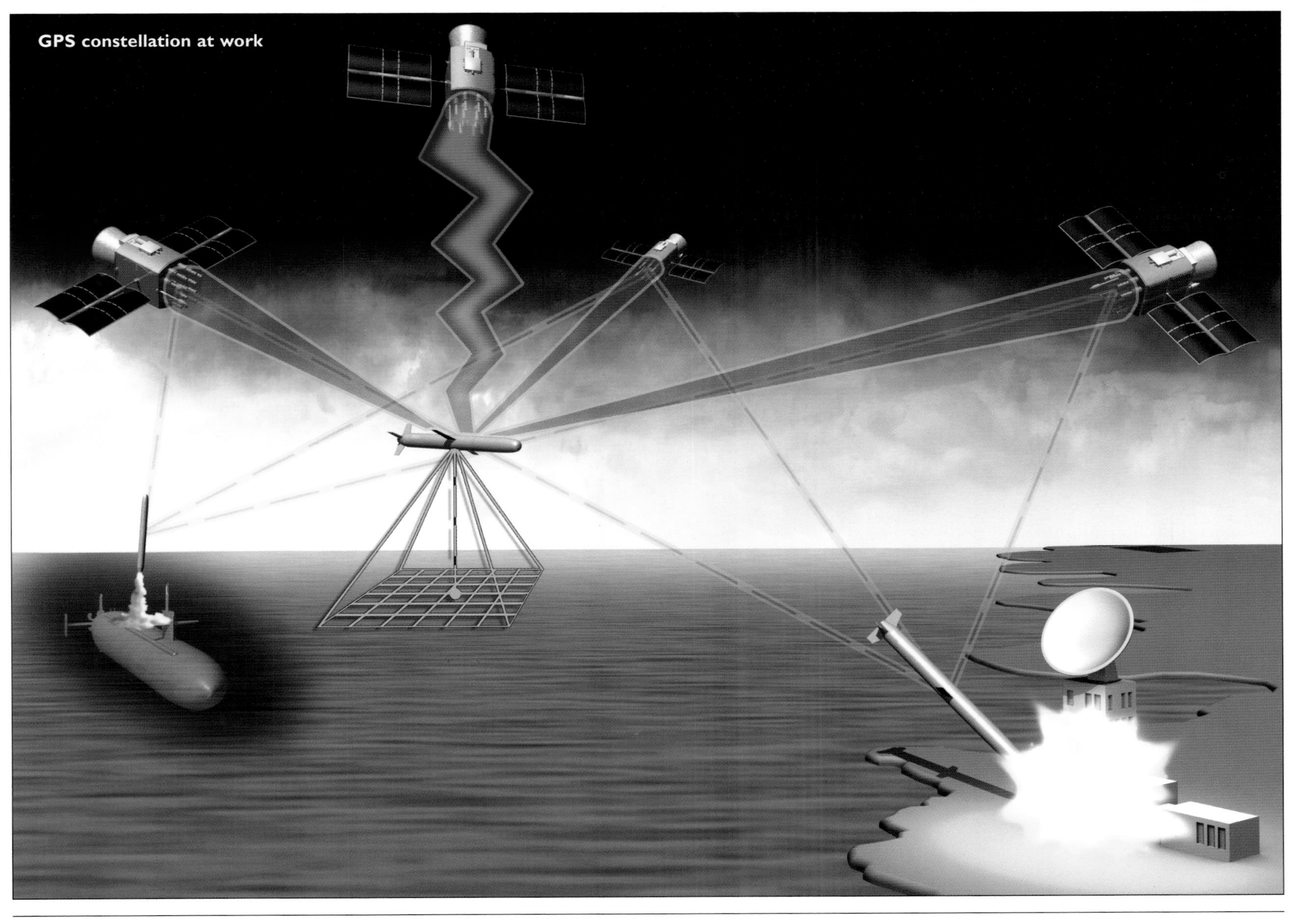
GPS constellation at work

SLVs can carry a large single payload or several smaller ones depending on their size, shape and orbit. This Titan III placed these early IDCSP communications satellites into orbit in 1968 with the aid of a dispenser. These satellites supported American efforts in the Vietnam War. (USAF)

velocity. This capability gives GPS unlimited application for military and civilian navigational uses. The satellite consists of four atomic clocks and a communications system to broadcast radio to a receiver unit between 65 to 85 milliseconds. The satellite has an unencrypted signal for most civil and commercial uses and an encrypted one for military uses.

The USAF conceived of an initial 16-satellite GPS constellation, which expanded to 21 satellites with three on-orbit satellites acting as spares. In 2004, there were 28 satellites in the constellation. Each satellite orbits the Earth every 12 hours and has an original service life of about seven and a half years, though the latest GPS models can serve for up to 12 years. Delta II boosters placed these satellites at regular intervals in an orbit of about 19,000km. Personnel from the 50th Space Wing at Schriever AFB operate the GPS and six ground monitor stations and four antennas control it. GPS horizontal accuracies range from 5 to 10m for military applications.

Today, GPS provides a significant role in navigational and position-reliant activities. Aircraft and surface navigational systems depend heavily on GPS to conduct operations in all weather, at night, over water and in other situations that rely on extreme accuracy. Scientist and engineers have also designed precision-guided munitions to exploit this accuracy. During Operation *Iraqi Freedom*, about 25 percent of 29,199 bombs and missiles used by Coalition forces were GPS-guided Joint Direct Attack Munition (JDAM) weapons. Pilots delivered approximately 5,500 of these weapons to within 3m of their targets.

Military leaders have always demanded reliable, fast and accurate capabilities to contact and receive information from fielded forces. Satellites provide instant, worldwide communications without having to rely on extensive cabling. As the military's need for long-haul, tactical and strategic communications overwhelmed existing telephone, telegraph and radio services, space communications came into being.

Today, the military relies on a series of communications satellite programs. The first attempts to provide a satellite communications network included the Initial Defense Communications Satellite Program (IDCSP). IDCSP used 45kg satellites launched in groups of satellites by a single Titan III booster. IDSCP was composed of 26 satellites and became operational in 1968 after four launches. This program provided support to Southeast Asian operations by providing voice and teletype support to commanders. Later, the Air Force Satellite Communications System (AFSATCOM) gave nuclear-capable forces a secure command and control system. AFSATCOM used the Navy's Fleet Satellite Communications System (FLTSATCOM) and the Satellite Data System (SDS) to send pre-formatted emergency war orders to bombers, ballistic missile silos and submarines. FLTSATCOM components in equatorial orbits and SDS satellites in polar orbit provided wide coverage for command and control purposes.

One of the longest serving systems is the Defense Satellite Communications System (DSCS). The current DSCS III, first launched in 1982, can provide commanders with a nuclear-hardened, jam-resistant communications link. The system has a constellation of five active satellites orbiting at 35,000km. AFSPC operates DSCS and, in 2004, it controlled 13 satellites, many inactive. Engineers are designing a replacement for DSCS III that will improve capacity, enhance Internet protocols, use laser crosslinks and improve aircraft and mobile ground communications.

AFSPC operates the joint service Military Satellite Communications System (MILSTAR) that gives users a secure, jam-resistant capability. The system has five MILSTAR satellites in GEO. Advanced technology has allowed MILSTAR's replacement, the Advanced Extremely High Frequency Satellite Communications System (AEHF), to provide five times the capacity in a smaller satellite and will operate from three to five satellites in GEO. The Navy also operates a number of communications satellites like its Polar Military Satellite Communications (Polar MILSATCOM) in an elliptical orbit. The Polar MILSATCOM program includes a three-satellite constellation, with each satellite weighing only 213kg. Naval commanders can also use the UHF Follow-On Satellite (UFO) that has largely replaced FLTSATCOM satellites developed for tactical use. First launched in 1993, UFO gives commanders ultra and extremely high frequency that produces jam-proof and secure communications. UFO has four primary and four back-up satellites in GEO. Another Navy system is the Global Broadcast System (GBS) that provides wide bandwidth communications capability to allow digital imagery and video to be sent to a tactical commander. The system consists of three GBS satellites in GEO and commercial back-up systems.

ABOVE The DSCS satellites gave commanders an anti-jam, secure communications capability. Voice and data transmission requirements have increased tremendously. Space-based communications have spread with the explosion of commercial systems. Today, the US needs military and commercial communications satellites to conduct operations. (USAF)

Although the United States has a wide array of military communications satellites, demand for services continues to outstrip capacity. Everyday communications, especially secure, long-haul circuits, are in continual use, especially for globally deployed forces. Like the public, the military has turned to the commercial market to enhance its communications capabilities with leased lines. Today commanders must consider commercial along with military communications satellites as potential enemy objectives to be destroyed or disabled by a potential enemy.

One of Eisenhower's original motivations in advancing a military space program was to acquire a means to avoid a nuclear surprise attack. The Air Force launch of the MIDAS program provided some success, but a ballistic missile early warning system needed better accuracy and reliability, especially if a president used this information to initiate a nuclear retaliatory

BELOW Satellites require extensive ground support to operate. Ground stations, like this DSCS one, can transmit and receive satellite signals. Their vulnerability to attack also makes them potential targets to disrupt space operations. (USAF)

Anti-satellite confrontation

LEFT **Anti-satellite confrontation**
Satellites have become a vital component for many nations. Military operations frequently need instant information from communications to navigation to guide force employment. Countries may counter this space advantage by using ASAT devices to destroy or disable the systems. This DMSP weather satellite is under attack by a direct energy weapon over Taiwan. Without current weather information, commanders might delay activities like aircraft strikes.

attack. The Pentagon created the Defense Support Program (DSP) in 1967. The concept was to place three to four DSP satellites into GEO that would identify ballistic missile launch plumes by a system of infrared sensors. Plume characteristics, collated data and other factors allow analysts to calculate the type of ballistic missile launched, when and where it came from and a potential impact point. This data is invaluable for civil and military defensive actions. Several countries now possess ballistic missiles and weapons of mass destruction that can be employed in a tactical or strategic role. DSP can detect both.

During the 1991 Operation *Desert Storm*, DSP gave commanders vital information about Iraqi SCUD and Al Husayn (a modified enhanced-range SCUD) ballistic missile launches, which allowed field commanders to activate Patriot anti-theater ballistic missile (ATBM) systems to down incoming theater ballistic missiles. Instead of a purely early warning system, these satellites have acquired a new role: ballistic missile defense tracking and guidance. Future systems will be designed to provide the ability to identify, track, and potentially guide antiballistic missile (ABM) systems to their targets. The USAF used the Titan III and IV and the Space Shuttle to lift these satellites into orbit. Technology has also enabled the DSP to conduct nuclear surveillance in order to detect unauthorized development and testing. A follow-on program to DSP is the Space-Based Infrared System. Engineers designed it for early warning and missile defense tracking and guidance.

Reliance on airpower to conduct strategic and tactical attacks has grown in conjunction with the improved speed, accuracy and stealth of the aircraft involved. Aerial operations depend on weather reports that can affect flight planning. Similarly, ground and naval operations also need current weather information. And weather satellites have been a key element of military space programs since its inception. The Defense Meteorological Satellite Program (DMSP) started operations in 1962. DMSP's original mission was to support photoreconnaissance satellite operations reporting conditions over the Soviet Union and other countries. Two satellites, in polar orbit at a 925km altitude, provide a range of environmental monitoring data. Growing civilian demands for weather information forced the merger of control of DMSP to a combined NASA, Department of Defense (DOD) and National Oceanic and Atmospheric Administration (NOAA) program office. NOAA has operational control of DMSP, but an Air Force Reserve unit maintains a back-up operations center.

Current military activities demand current and intelligence data to plan, prepare and execute operations that can affect activities from updating cruise missile targeting to monitoring enemy communications. These capabilities are under the control of United States intelligence agencies. Reconnaissance and surveillance functions include photoreconnaissance, electronic signals interception, radar imagery, and other functions. Although linked to DOD, agencies other than the USAF operate these satellites.

Launch crews sent GPS satellites into orbit by the tried and tested Delta II SLV. These vehicles used technology from the Thor intermediate range ballistic missile. Solid rocket boosters around the liquid-fuel first stage provide additional lift capability. (DOD)

RIGHT Although satellite operations are relatively self-contained on the spacecraft, ground support is still required. This ground station at Kaena Point, Hawaii, allows controllers to communicate with the satellites, assess their health and order changes to their operations. (USAF)

Fighting in and from space

A more controversial use of space systems involves deploying weapons from space or their use within space. Most countries do not consider devices such as ballistic missiles that traverse through space and land on the earth's surface as space weapons. However, at a minimum, nations agree that deployed satellites that act as space mines, platforms to deploy munitions, ASATs or other devices that can disable or destroy another space or ground system, are space weapons. Some critics argue that ground-based systems that can attack or effect space systems are "space" weapons. Militarization of space has occurred since the late 1950s, but the weaponizing of space has raised many new issues.

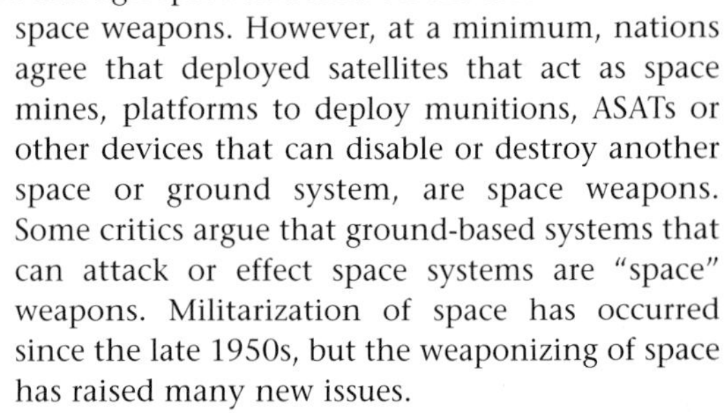

The two general categories of space weapons are: ASATs and space-deployed weapons systems. ASAT requirements and capabilities are not new. The United States operated two ground-based ASATs during the early 1960s developed from existing programs. The first was an evolutionary development from the Nike family of surface-to-air missiles. The US Army had designed and deployed a continental air defense system to protect the nation against Soviet bombers. When the major strategic threat came from ballistic missiles the Army responded by building Nike-Zeus as an ABM. Although earlier Nike missiles could destroy an aircraft with a high explosive warhead, the Army armed later systems, like Nike-Zeus, with a small-yield nuclear weapon. Advances in ballistic missile technology made ABM operations even more difficult due to high speeds, multiple reentry

BELOW Space and missile development evolved together. Early ballistic missiles found a home as SLVs. The USAF never fired the Thor ballistic missile, armed with a 1.44-megaton warhead, in anger. However, it served as an SLV and nuclear ASAT delivery vehicle for years. (USAF)

The USAF's Program 437 could destroy a satellite in LEO with a nuclear weapon. This control center would launch the Thor from Johnston Island south of Hawaii. Crews relied on primitive computers to calculate targeting information. (USAF)

vehicles (RVs), decoys and reduced warning times. Nuclear weapons increased the lethal destruction area against incoming RVs and improved the probability of destroying the system. Similarly, a nuclear-armed Nike-Zeus could destroy an orbiting satellite given the great speeds and imprecise orbital calculations at the time. Army crews operated a single Nike-Zeus, armed with a 400-kiloton nuclear yield, at Kwajalein Atoll in the Pacific. Program 505 "Mudflap" became operational on June 27, 1963. A single missile would have had limited ASAT capability.

Washington replaced the Nike-Zeus system with a 1.44-megaton-yield Thor ballistic missile. USAF crews could launch two Thor missiles within hours of an alert, though tracking a satellite and ensuring the liquid-fueled missile was ready to intercept it was a challenge. These same Thor missiles, stationed on Johnson Island south of Hawaii, had served as high-altitude nuclear test devices. During one test on July 9, 1962, an Air Force crew detonated a nuclear device at an altitude of 400km that created sufficient electromagnetic pulse (EMP) to cripple three orbiting satellites, including a Transit navigational satellite, by damaging their solar panels and electronic components. It also affected electrical grids in Honolulu. This test demonstrated the powerful capability of an ASAT that could deliver a nuclear strike on an orbiting satellite within an 8km kill radius. Under Program 437, USAF launch and maintenance personnel from the Aerospace Defense Command (ADC) operated the system from May 28, 1964, to October 2, 1970. The cost, slow response, drain of the Vietnam War on personnel, and expansion of more space targets than ASATs caused the USAF to abandon the project.

ADC launch crews operated two nuclear-armed Thor missiles as ASATs. Although crews could launch their missiles against a satellite, they had limited logistical support and were subject to many concerns from security to weather problems. (USAF)

RIGHT **Space weaponization**
The Air Force has dreamed of using hypersonic space vehicles to transport systems into space. In the late 1950s, an experimental hypersonic glide vehicle, the X-20 Dyna-Soar, would have explored the possibilities of such requirements. Dyna-Soar could have extended strategic bombardment from space. This hypothetical capability would have given the United States a great advantage since there was no defense against this type of attack or long-range warning system in place to detect one.

Other potential ground-based weapons include directed energy weapons (DEWs). There are several nations capable of focusing electromagnetic energy or neutral particle beams to destroy, disable, disrupt or deny space systems or their operations. The United States, Russia and China have the ability to use high-energy lasers and these nations have all tested DEW systems as ASATs. These weapons allow rapid targeting, reuse and precise targeting. A laser could destroy a component, through overheating, that damages the satellite so that it cannot function. It also avoids collateral damage or debris after the target has been disabled. A laser operator would need to ensure the weapon remained on the target long enough to build up sufficient heat while the satellite moves at great speed. The USAF is developing an airborne laser for use in ballistic missile defense that could potentially become a modified ASAT.

Satellite operations depend on ground operations to receive instructions and transmit data. If those communications links are disrupted or negated then the satellite cannot operate so electronic warfare could be used to affect national or international systems without having to go into space. Many countries could develop the ability to jam signals creating a temporary disruption of service. Additionally, selective signals could be jammed allowing for a limited service of certain functions. False signals could also be sent to trigger a shutdown of a system or make unexpected maneuvers to a different orbit. If the electromagnetic signal has ample power, then it could harm an electronic component. Electronic warfare capability is widespread and a number of states have airborne, naval and ground-based systems.

The USAF continued to seek ways to conduct ASAT missions in the 1970s. One approach was to use high-speed fighter aircraft to deliver non-nuclear kinetic weapons that would strike a satellite and obliterate it. The aircraft would fly to

The Soviet Union tested and developed a co-orbital ASAT that could destroy a satellite by creating a field of debris to disable fragile spacecraft components like solar arrays. Although the Soviets maintained the system, it was never used. (Department of Defense)

Space weaponization

The F-15 equipped with a miniature homing vehicle ASAT required precise guidance to destroy its satellite target. The vehicle used components from existing systems that reduced cost and increased the probability of successful interception. The program never reached operational status. (USAF)

its maximum altitude and release the weapon after which a rocket with an internal homing device would guide the kinetic kill system to its target. The booster system had to have sufficient speed to catch the target while an accurate tracking system was needed to send the aircraft to an appropriate intercept point. USAF engineers experimented with F-106 interceptors to deliver a modified two-stage Standard anti-radiation missile. By 1975, President Gerald Ford authorized the USAF to build an F-15 interceptor-launched kinetic vehicle that used a Short Range Attack Missile as a first stage and an Altair III as a second stage. This miniature homing vehicle (MHV) was tested on September 13, 1985, against a Solwind satellite, which was destroyed at an altitude of 515km. The Air Force planned to arm F-15 squadrons at Langley AFB, Virginia, and McChord AFB, Washington, with the MHV but fears of a potential space arms race forced the cancellation of the project in March 1988.

ASAT operations are not limited to ground-based or aerial interceptors. The USAF investigated the use of an orbital satellite inspector under Project SAINT in the late 1950s. This satellite was designed to attain orbit, intercept and inspect a space system; it was never put into production. However, the Soviet Union did develop and test its Polyot co-orbital ASAT device. Aware of CORONA and other intelligence gathering satellites, Moscow constructed an ASAT device that would track, intercept the satellite within one to two orbits and explode a device that would send fragments towards the satellite, destroying it. The Soviets launched two tests in October and November 1968 that demonstrated this concept and test flights continued up to December 1971 when a test ASAT intercepted its intended objective, another Soviet satellite in a 250km circular orbit. Some American officials believed the Soviets could intercept satellites at an altitude of up to 4,800km. The program ended in 1982. Mutual concerns about a space arms race and international pressure helped Washington and Moscow sign the 1967 Outer Space Treaty prohibiting military bases on the Moon and other celestial bodies. The agreement also banned the placement of nuclear weapons in orbit, reducing fears of MOBS programs or nuclear-armed co-orbiting ASATs.

A country could use several systems to deploy weapons from space. Ballistic missile defense platforms or bombardment systems seem plausible. These systems could include unmanned satellites or a manned system, like a space station or hypersonic vehicle. The United States did propose a number of systems that critics considered as space weapons. Under President Ronald Reagan the nation explored a space-based ballistic missile defense shield. In a speech on March 23, 1983, Reagan initiated his Strategic Defense Initiative (SDI, sometimes referred to as his "Star Wars" program). The program examined systems such as orbiting DEW platforms, satellites containing kinetic interceptors and reflecting dishes for ground-based DEWs. DEWs require large quantities of energy while the use of chemical lasers in space is limited due to fuel storage. Reflecting ground-based DEWs with a space mirror could circumvent these concerns. SDI engineers also explored the use of nuclear weapons, though if the US had deployed these systems then it would have had to renege or renegotiate several arms control agreements, including the 1967 Outer Space Treaty and the 1972 ABM Treaty that limited ballistic missile defenses. Fears of Soviet ICBM capability and a possible Moscow-controlled DEW orbiting space system motivated Reagan to press for SDI. With the collapse of the Soviet Union and, as cost and technical issues mounted, policymakers replaced SDI's emphasis on strategic with theater ballistic missile defenses. Today, there are calls for additional space-based ballistic missile defenses due to the growth of threats from several nations that have acquired SCUDs and longer-range missiles.

F-15 pilots conducted several tests of an air-deliverable ASAT weapon. Although the Air Force never deployed the system, Washington had developed a weapon capable of destroying a LEO satellite with kinetic effect. The Air Force would have armed two squadrons to conduct this mission. (USAF)

Countries can deploy several types of orbital bombardment systems. A state can use a MOBS that consists of a satellite placed in orbit that contains rockets armed with either conventional or nuclear weapons. If the country puts the MOBS in GEO, it could aim a weapon above a target indefinitely. After Sputnik, Washington was very concerned about a nuclear-armed MOBS stationed above the US. USAF ICBM efforts were floundering due to technical and cost concerns, and Soviet Premier Nikita Khrushchev fed these fears by proclaiming that he would orbit nuclear weapons over America. This had led to Washington deploying Programs 505 and 437. Another space system of concern is the fractional orbiting bombardment system (FOBS), which sends a ballistic missile in a lowered flight trajectory to its target from any direction.

The US explored the use of manned systems to wage combat in and from space from the 1950s and early 1960s. Technological advances from programs such as the X-15 provided the Air Force with the impetus to develop a manned

The Air Force wanted to develop and deploy the X-20 Dyna-Soar into space. The vehicle, launched by a Titan III, could have evolved into a weapon capable of conducting reconnaissance, ASAT and, potentially, strategic bombardment missions. Washington canceled the program due to cost, technical and political concerns. (USAF)

The X-15 explored hypersonic speed travel. X-15 crews set altitude records that qualified them as reaching space. These experiments convinced many Air Force leaders about manned space programs, but their participation was limited for various reasons until the advent of the Space Shuttle. (USAF)

space force. USAF personnel and contractors examined concepts to build an experimental, reusable space plane, the X-20 Dyna-Soar, which would become an operational system. Air Force launch crews would use a Titan IIIC to propel the X-20 to hypersonic speeds of Mach 5 to 25 to conduct reconnaissance, ASAT, satellite inspection, supply and repair of on-orbit space systems, and orbital bombardment with nuclear weapons. Fears of a Soviet clone coupled with the tremendous cost of the project forced its cancellation by Secretary of Defense McNamara in October 1963.

NASA and the USAF have continued to investigate options to provide manned hypersonic space flight to fulfill many of the roles designed for the X-20, though the lethality and precision of current conventional weapons has made their use preferable to that of nuclear weapons. Greater speed to conduct global strike missions, fewer overseas bases and greater interest in areas not easily accessible, technological capability, growing awareness of space's importance, and the fear of other nations gaining an advantage in space has led to extended interest in deploying and operating a space plane capable of carrying munitions around the globe in minutes.

The use of space in warfare has significantly changed from one of support for terrestrial operations to one where operations from space are starting to originate above the atmosphere. Although the US has led the way in the majority of operations, Russia and China have also examined the use of space weapons. As space assets become more vital to air, land and sea operations, space as a new environment for present and future operations will grow.

This modified Lockheed Martin TriStar carries the Pegasus three-stage solid-fuel SLV. The aircraft-deployed Pegasus can carry relatively small payloads into LEO. Pegasus provides the flexibility to launch satellites from many locations. (USAF)

Striking from the deep black: space systems go to war

For the United States space militarization began after World War II. The German V-2 demonstrated the potential to lift payloads into the outer edge of space. Satellites could orbit the world and provide new means to conduct traditional military actions. Driven initially by the desire to develop a new nuclear capability, Washington soon sought ways to protect itself from a similar potential Soviet threat. Space offered the ultimate high ground and seemed like a natural extension for exploitation. If the US could exploit these technologies, then it could provide new capabilities to defend itself in the nuclear age. Washington was concerned that the Soviet Union would counter its ability to use space. Diplomatic, technical and, finally, military options were considered to assure it could have access to its satellites. The country needed reconnaissance and surveillance satellites to ensure it could detect any buildup of Soviet bombers or ballistic missiles. Early warning satellites provided instant alert of a nuclear attack. Only after the launch of Sputnik did Washington seriously investigate developing ASAT weapons. Military space efforts have evolved as technology, perceived threats, political will and national interests have changed.

Cold War applications

The US military use of space was primarily to support operations. Communications, early warning, weather and various reconnaissance systems dominated space operations for strategic purposes. As the Cold War progressed, the focus of space systems started to support more tactical purposes. USAF engineers and scientists investigated many potential space weapons without ever deploying an orbiting platform. Program 437 was the primary ASAT capability, but Air Force crews could only launch two missiles and it soon became obsolete due to the proliferation of Soviet space and missile efforts. Manned USAF space efforts, like Dyna-Soar, were never realized, but became collaborative space research projects with NASA.

Throughout the early space program, launch failures exceeded successes. SLV booster failures, satellite malfunctions and other breakdowns confirmed that space systems were still highly experimental. However, the systems that did work provided vital strategic information. The Air Force never established a dedicated space corps or major command until 1982. Instead, the Air Force relegated space to its research and development arm to launch satellites and then turned them over to various operational commands like SAC or ADC. Military space became an orphan in search of a parent.

Throughout the Cold War the tremendous use of reconnaissance and early warning systems allowed the United States to develop strategies to combat the growth in Soviet strategic systems.

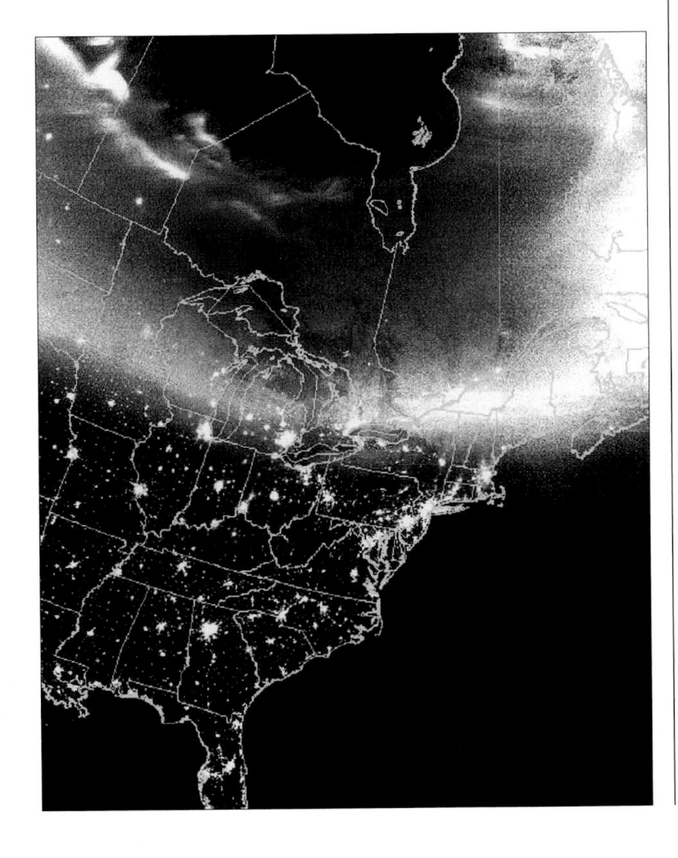

Weather reporting not only presents valuable information for planning, but also ongoing operations to recall aircraft or maneuver orders. Countries could use imagery like this from commercial sources and unrestricted public satellites to plan aerial operations and attacks. DMSP satellites can distinguish light concentrations easily and identify urban areas. (USAF)

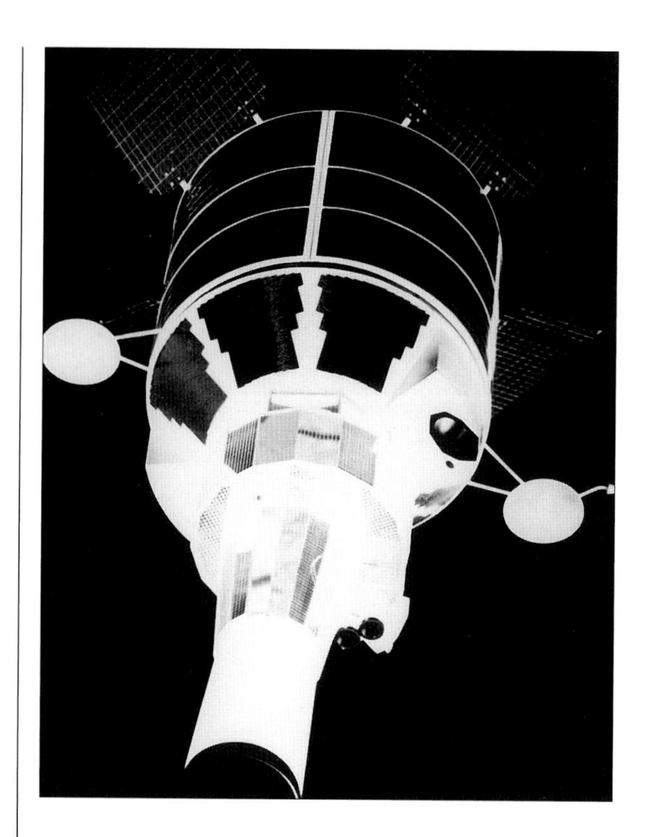

DSP was the successor to the MIDAS satellite family. DSP satellites gave the United States a powerful tool to avoid a nuclear attack. Early warning was vital to America's nuclear deterrent force. These satellites provided significant help in the 1991 Persian Gulf War. (USAF)

Deterrence strategy and arms control measures helped the country adjust to a nuclear peer competitor. Expanding use of space also gave national civilian and military leadership greater confidence through situational awareness that allowed the nation to conduct global operations. Communications, imagery, weather and other force support capabilities allowed the country to conduct activities without a huge military presence, which enabled quicker reaction times and reduced presence, and thus vulnerability.

As technology improved, greater opportunities presented themselves. GPS created options to improve traditional navigation for ships, planes and ground forces, and it also allowed the manufacture to build precision-guided munitions. Instead of using massive, inaccurate conventional bombing that put aircrews at risk and increased potential collateral damage to civilians, precision-guided munitions allowed fewer aircraft to conduct more accurate missions. Greater precision also meant that engineers could fit new weapons, such as cruise missiles, onto existing airframes, extending their use in service; bombers like the venerable B-52 found new lives as cruise missile carriers.

The Air Force also sought a wider mission away from its purely aerial role. Although space capabilities grew slowly and appeared more experimental than operational, combining the field of aerial and space operations created new possibilities. Conversely, the rise in satellite reconnaissance capability and the transfer of its ownership to non-DOD intelligence agencies allowed the CIA to dominate the control of information, even though the USAF had personnel in the NRO. Although this process integrated disparate intelligence-gathering organizations, it also limited access to the data by military planning and operating forces. It also restricted space reconnaissance efforts to strategic views, whereas military commanders required operational or tactical levels of information. Rivalries and protection of access may have limited any potential gains from a centralized use of space.

The 1991 Persian Gulf War

By 1990, the end of the Cold War was near and the US was left standing as the world's sole superpower. Geared towards fighting a conventional war in Europe against the Warsaw Pact, US military forces trained, organized and were equipped for action in Central Europe. Space systems were developed to maintain surveillance and conduct operations against Soviet aggression around the globe. Systems designed to watch Soviet military movements and operate under nuclear conditions seemed obsolete. However, global communications, weather, navigation and other functions that allowed forces to operate globally were very much relevant as America faced a "New World Order." Despite dreams of "peace dividends" and thoughts of an end to conflict, the world started to erupt in regional hostilities. Historic problems concerning geographic, ethnic and religious hatreds boiled up to major conflict. Instead of fighting in Germany, the US would deploy forces in areas where there were no support bases, in difficult terrain and facing unknown foes.

One of the most serious regional outbreaks occurred between Iraq and Kuwait in 1991. Iraq had already fought a war with its neighbor Iran over a territorial dispute. After several years of conflict, Iraqi and Iranian forces had settled into a static, World War I-style trench war, though they had traded limited ballistic missile volleys. Iraq had also used modified Soviet-era SCUD missiles to extend their range to strike Tehran during the "War of the Cities" from February to August 1988, when both sides traded attacks on their respective capitals to weaken the will of the people. Iran's public will started to crumble based on the continual attacks, lack of resources due to economic sanctions and failures on the battlefield that eventually forced them to sue for a ceasefire.

Saddam Hussein, despite the Iraqi victory, had depleted his nation's resources to fight the war. Saudi Arabia and other neighboring countries, like Kuwait, had provided loans to Iraq to contain Iran. Fearful of allowing the spread of radical Islamic fundamentalism that would topple their regimes, oil-rich Persian Gulf potentates supported Hussein. Saddam believed that he had repaid his debt in blood and wanted more aid to rebuild his country. He also had expressed previously the idea that Kuwait had stolen oil from reserves near its border and that historically the nation was part of Iraq. The upshot was that Iraq invaded Kuwait on August 2, 1990.

The Persian Gulf War would later become the first real application of space systems in combat. America's military forces had started to integrate several space systems into its land, sea and air forces. For example, aircraft had GPS receivers, though only a few units had receivers in 1990. The USAF had equipped only 72 F-16C/Ds, 37 B-52s, 21 RC-135s and a handful of other aircraft with GPS. The other services had fewer still, with the US Marine Corps not employing any GPS receivers in their aircraft. Space systems were already on orbit and ready to use, though they would need to be repositioned, equipment purchased and forces deployed in theater to support operations. However, the Coalition and Iraqi military leaders would soon discover the advantage of instant and reliable information from military and commercial space systems. Information from communications to early warning satellites would dominate military decisions and planning.

American military forces started to arrive in Saudi Arabia shortly after Iraqi forces took over Kuwait. The US government was concerned about Iraqi armored columns turning south into Saudi Arabia and other Gulf states. Iraqi control over Saudi Arabia would give Hussein the largest oil reserves in the world and could ignite global political and economic chaos. Historically, Washington, concerned

The US first used space communications extensively during the 1991 Persian Gulf War. Use of satellite terminals became an instant hit with field commanders as it allowed them immediate access to information. (DOD)

DSP and satellite communications allowed Patriot missile batteries to intercept Iraqi ballistic missiles. Early warning by DSP allowed missile defense crews to shoot down incoming SCUDs. (DOD)

about Persian Gulf issues, had depended on Iran and Saudi Arabia as sources of stability. However, the Shah of Iran had fallen and been replaced with a revolutionary government hostile to the West. Now Iraq threatened Saudi Arabia. American military forces within the area under Central Command (CENTCOM—the US regional combatant command responsible for planning and executing military operations in the Middle East) were relatively small compared to the European or Pacific Commands.

CENTCOM operations and planning staffs required instant communications from units deployed in the desert who did not have access to land-line telephones; radio was also unable to handle communications with service components stationed from Georgia to Hawaii back to the Gulf. Over 500,000 US military personnel in the Persian Gulf would need communications as would European forces deployed to the Gulf. Similarly, Coalition allies from around the world reacted to United Nations resolutions that demanded Iraq return Kuwaiti sovereignty and eventually authorized member nations to use force to do so. Communications capabilities needed instant support; the Pentagon threw DSCS and others into the fight. Ultimately, ten different communications satellite systems would carry over 90 percent of US communications in the Gulf. Commercial satellites supported about 24 percent of the requirement.

Throughout Operations *Desert Shield*, the preparation leading to combat operations, and *Desert Storm*, combat operations, DSCS and other systems gave commanders key command, control and intelligence information. AFSPC had to reposition a DSCS satellite and use spare systems. The Pentagon also had to lease commercial systems and take other actions to resolve bandwidth problems. Washington also asked London to use a British Skynet IV-B satellite in the area for additional capacity. DSCS shouldered 75 percent of all intertheater connections during *Desert Shield*. Ultimately, DSCS would carry up to 84 percent of all long-distance strategic communications during the conflict. CENTCOM had only four tactical DSCS tactical terminals in the area before the Kuwait invasion. Before combat operations commenced on January 16, 1991, there were 120 DSCS terminals around staffs deployed throughout the Persian Gulf allowing commanders to control combat actions, provide support or coordinate joint and multinational service activities. Eventually, CENTCOM staffs would use about 7,200 terminals working with 63 different satellites that included communications, intelligence, navigation and weather systems. For example, air operations in the theater relied on a massive air tasking order (ATO) that specified target, support, and other information for aircraft sorties. On the first night of air operations over 950 aircraft conducting flying missions, while on most days Coalition air forces flew up to 2,500 sorties daily.

The Air Force and other forces were concerned about space communications. Questions arose about Iraqi capability to jam satellite communications links. The US had a handful of secure satellite communications circuits that included FLTSATCOM and AFSATCOM. However, many circuits like leased commercial lines had more capacity than DSCS or other military-controlled systems. The transponders used by these commercial systems, e.g. INTELSAT, used relatively low-power transmissions that were susceptible to jamming. Fortunately, no one had leased the circuits and the United States had sufficient financial resources to use these commercial lines and Iraq had insufficient means to jam the signals. The United States also had to employ communications satellites that were well beyond their useful service lives. If one failed, the nation did not have any capacity to immediately launch a replacement satellite.

The Patriot missile defense system was based on a surface-to-air missile design. The integration of this tactical system and the strategic DSP system allowed commanders to defend several areas around the Middle East. (DOD)

One of the largest challenges for military operations is to maneuver into a proper location and fire on the correct target. Both tasks require precise navigation and locational information—a particularly difficult task in the deserts of the Persian Gulf. To complicate matters further, military units performed much maneuvering and many other military operations at night. The use of the GPS system enabled accurate navigation for anyone equipped with a receiver.

Demand quickly outstripped the supply of military receivers and Coalition military forces had to procure commercial receivers. By March 1991, military users employed 4,490 commercial GPS receiver units compared to 842 units designed to use military signals. The Air Force GPS program office would eventually buy over 13,000 civilian GPS receivers for military vehicles. This condition forced the Air Force not to implement GPS selective availability that would have inhibited non-military reception of its signals. Although limiting commercial GPS use would have hurt transportation activities worldwide, it would also have limited any potential Iraqi use of the system. However, the Coalition forces needed GPS to maneuver, operate in the trackless desert and enable navigational systems to guide missiles. Coalition military leadership judged Iraqi GPS use as minor at best, since their ground forces had entrenched in static defensive positions in Kuwait. Still, the military receiver units allowed personnel to get within a minimum of 16m of a true position; commercial users could be assured of accuracy within 25m.

The US had not fully deployed its GPS constellation due to the *Challenger* disaster and a shortage of SLVs. Instead, the USAF had to use existing test satellites for operational use. Instead of 21 operational and three spare GPS satellites, military users had to rely on only 16 satellites in total. Despite these shortcomings, GPS enabled military forces to conduct many actions. Early GPS models had markedly improved precision compared to other systems. During a USAF test seven years before Operation *Desert Storm*, a GPS-equipped F-4 bombed targets at 3,000m with a circular error probable (CEP) of 20–30m. CEP is a measure of relative accuracy that measures the radius of a circle in which half of all munitions launched at a target expect to fall. Just before hostilities

GPS was another innovation used widely for the first time in direct combat. GPS allowed navigation in the featureless theater of operations. Tanks could traverse long stretches of desert without getting lost. (DOD)

started, CEP fell to only 10m due to improved training. Coalition military forces were able to use GPS to launch Tomahawk Land Attack Missiles by getting precise launch coordinates for a ship launch. Ground forces maneuvered up to 2,000 vehicles in its land offensive over a 120km front. Naval forces could clear, lay and navigate through minefields. Soldiers could target enemy forces with artillery fire and close air support. Aircraft could navigate at night; they also delivered munitions with precision to support ground forces.

Desert Storm military planning staffs had concentrated on using an extensive air campaign to isolate and disable key Iraqi government and military leadership activities. They also attacked communications networks, weapons of mass destruction sites, ballistic missile targets, transportation nets and electrical grids. Flying in the desert was difficult, but GPS aided this activity. However, weather conditions in January 1991 were the worst in 14 years: unusual weather patterns including heavy rain, sandstorms and fog contributed to flying problems. Heavy rains could turn desert areas into instant swamps and certain areas had zero visibility due to sandstorms. Weather changed within minutes. In one instance on January 24, DMSP indicated clear weather over Baghdad while Basra, on the coast, had overcast conditions. Within 90 minutes, the conditions reversed. Oil well fires also obscured flying activities. DMSP and other commercial environmental monitoring satellites could aid aerial and other military operations.

DMSP and remote sensing satellites allowed military commanders to plan and schedule actions that required relatively clear weather. Reconnaissance, attack sorties employing laser-, infrared- and optical-guided munitions, bomb-damage assessments, retargeting aircraft and other actions that needed cloudless weather required DMSP. Ground commanders also used DMSP to plan maneuver routes. All services used environmental monitoring satellite data to analyze terrain. LANDSAT and other commercial satellites with multi-spectral capabilities allowed air staffs to evaluate terrain and plan attack routes. For example, USAF F-111 fighter-bombers sealed burning oil well heads at the Mina al Ahmadi complex in Kuwait by using this data. Hussein's retreating troops had left Kuwait by burning oil wells or creating spills in an attempt to delay the Coalition advance and destroy Kuwaiti assets. F-111 crews required detailed bombing procedures to close the Mina al Ahmadi oil well heads. DMSP also provided bomb-damage assessment on Iraqi electrical grids by measuring electrical lighting and other targets during the first night of a strategic bombardment on January 16. DMSP allowed air commanders to cancel sorties over targets that had poor weather conditions via ATO modifications. Military leaders, using DMSP data, canceled flights that saved up to $250 million with sorties that would have returned home due to bad weather anyway.

Multi-spectral satellites gather information by taking imagery with different frequencies. These images provide information that is not available to human eyes. LANDSAT and the French SPOT satellites also provided data to commanders not on published maps. Many published maps were 10–30 years old. LANDSAT and other imagery provided sufficient data to create new maps. They also gave commanders information about subsurface features. Unfortunately, LANDSAT was only available over the area every 14 days and the Coalition needed commercial multi-spectral satellites to supplement it.

These satellites provided many examples of their worth. During planning for a possible airborne invasion of Kuwait City, the 82nd Airborne Division requested information about possible obstacles and any defensive measures that would inhibit this action. Similarly, LANDSAT and SPOT imagery indicated areas that appeared capable of supporting armor operations, but were sabkhats—areas that contain water covered by a thin surface crust, a natural tank trap. Tanks or armored vehicles would sink if commanders traversed this area. Navy and Marine Corps staffs asked for information about coastal and open sea areas. Multi-spectral data gave information about conditions 10m below the surface that could inhibit an amphibious invasion.

Coalition forces started planning for the use of ballistic missile attacks by Saddam Hussein on civilian and military targets. Iraq had used chemical weapons in the past against the Kurds, as well as using SCUD and other ballistic missiles against civilians in its Iran–Iraq War in the 1980s. Iraqi missile crews had fixed and mobile SCUD ballistic missiles with ranges of up to 300km. Their flight time was about seven and a half minutes, not much margin of error to detect or intercept these missiles. Baghdad had also modified these SCUD missiles to strike longer-range targets, like Tehran, during the "War of the Cities" in 1988. These Al Husayn missiles sacrificed payload for a longer range, but they could still carry a chemical or biological weapon and strike into the heart of Saudi Arabia or a military target without the requirement for great accuracy to produce a terror attack. Potential actions to defend sensitive targets included deploying Patriot missiles, issuing chemical suits to personnel and using DSP satellites.

Space systems allowed field commanders unprecedented, real-time information. Navigation and communications allowed aircraft to conduct precision strikes like these on the Ahmer Al Jaber airfield hangars used by Iraqi forces. (DOD)

DSP satellites were originally not a part of any Iraqi ballistic missile warning system; they were still assigned as an element of a vast strategic early warning system geared towards the Soviet Union. Air Force DSP satellite operators had never had to detect and provide warning against SCUD or tactical ballistic missiles before, though they had monitored launches in the past. AFSPC DSP crews had observed over 600 of these launches during conflicts in Afghanistan, the Iran–Iraq War and test launches before the Persian Gulf War. These crews had sufficient information about their launch characteristics to analyze some SCUD activities. Saddam Hussein had unexpectedly fired three test SCUD missiles towards Israel. Perhaps he used these launches to scare Israel into a retaliatory attack or adjust his ballistic missile targeting to hit Jerusalem. AFSPC crews detected the launches, but found their procedures lacking and needed to improve their ability to reduce warning time. AFSPC DSP crews also had to maintain their main mission of protecting the nation from a nuclear threat. They had additional missions to detect space launches and nuclear testing, not tactical ballistic missile launches. Although CENTCOM headquarters knew about Iraqi ballistic missiles, most planners initially dismissed them as nuisances.

Political and military commanders soon realized that although the SCUDs had a relatively small warhead and were very inaccurate, they had a vital strategic potential to alter the war. If Hussein could launch missiles against Israel, then the spectacle of Muslim countries fighting against a fellow Islamic nation whilst protecting Israel might create problems. The effort to stem Iraq might disintegrate if Tel Aviv responded to a SCUD launch with an attack on Baghdad. Additionally, if an Iraqi SCUD had used a chemical weapon and hit any civilian population center, Israeli or not, then its detonation would have immediate repercussions. Patriot ATBM units moved into area during *Desert Shield* and more units joined existing ones soon after *Desert Storm*. Space experts had proposed using DSP to improve Patriot operations. Providing more time to the Patriot's launch crews would aid interception by allowing the interceptor missile to catch an incoming SCUD or Al Husayn earlier in flight. A Patriot battery could also fire more than one ATBM weapon against a target given more warning time.

Typically, Patriot operations would begin with an Iraqi SCUD or Al Husayn launch. A number of sources could detect the launch: a fighter patrolling in the area, airborne radar aircraft or DSP. In August 1990, the Air Force had three DSP operational satellites and two spares in GEO. The Air Force operators knew DSP could work for SCUD and Al Husayn missile warning. AFSPC had to conduct

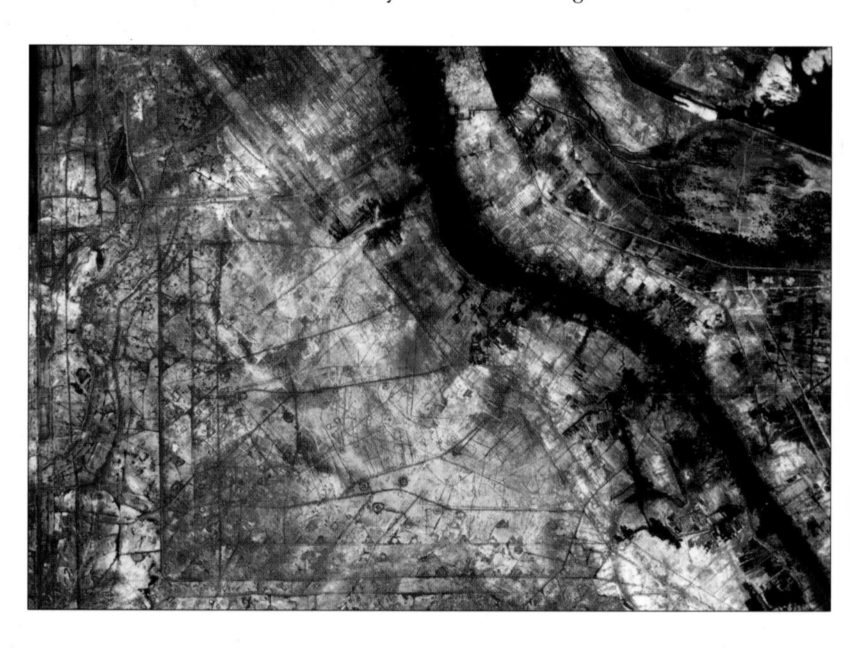

Remote-sensing satellites can offer a commander invaluable information. This LANDSAT image from Iraq can give military staffs data on terrain conditions such as this wetland area in Iraq. (NOAA)

The North American Aerospace Defense Command collects space-based information to assess missile and air threats to the United States and Canada. Information from DSP satellites allows controllers like these in the operations center to notify national leadership of an attack. (NORAD)

the strategic warning mission while using the two DSP spare satellites for the Persian Gulf and one of the active satellites. This entailed moving satellites into proper orbit, a difficult maneuver given the limited amount of onboard fuel.

Analysts then relayed the appropriate detection, after study, to CENTCOM. CENTCOM then advised an appropriate response to Patriot missile batteries for incoming warheads. CENTCOM aircraft also started searching for the mobile transporter erector launcher vehicles carrying SCUDS. These warnings also triggered warnings and civil defense measures in cities.

DSP satellites have an infrared telescope that can track the earth every ten seconds for a launch plume, though weather patterns, such as clouds or humidity, can slow down the detection. If a satellite detected a possible Iraqi launch plume, then satellite operators sent that information through a ground station and then transmitted to a satellite control facility at the Buckley Air National Guard Base and other stations in Europe. Concern about Iraqi agents forced AFSPC to move six mobile DSP satellite-processing vans to locations where they could support operations if the terrorists disabled any ground stations. Ground station crews forwarded the satellite data to the North American Aerospace Defense Command (NORAD) facility in the Cheyenne Mountain Complex near Colorado Springs. NORAD is responsible for strategic missile and aerial warning to protect the US and Canada and had the experienced staff and resources to analyze the satellite data to determine if the launch information was valid. NORAD officials also had to segregate the theater ballistic missile threat from its main objective of protecting the US and Canada from an ICBM or SLBM attack. Despite the lack of Iraqi ICBM capability, an independent focus on national ballistic missile warning had to maintain a credible vigilance of Russia and China. Theater ballistic missile warning required specialized training and detection techniques. SCUD and Al Husayn missiles had less powerful rocket engines than ICBMs or SLBMs. These missiles produced a smaller, less intense infrared exhaust signature that burned for less time than an ICBM and required a greater fidelity of detection. Specialized tracking was required for detection and analysis.

Once the information was processed, Space Command officials had to respond quickly to transmit the warning to CENTCOM. A ballistic missile warning had two ways to reach Patriot commanders in Saudi Arabia, Israel or Turkey: satellite communications through DSCS or other military systems was available to notify appropriate commanders, or NORAD or other ground stations could simply contact the CENTCOM staff through a telephone call. Information was also available through the Tactical Event Reporting System (TERS). NORAD analysts sent their warnings through DSCS into the TERS. Army, Navy and USAF units, including the CENTCOM staff, received the appropriate warnings through TERS to take action. CENTCOM officials, after getting the TERS data, would then send a warning via satellite communications, including DSCS, to the appropriate Patriot units. The Patriot battery commander would then initiate actions to search, track, target and intercept the incoming ballistic missile. If the Patriot battery used its own detection radar, then they typically had 90 seconds to detect, track, lock-on, intercept and destroy a SCUD. Patriot's radar had an 80km range; DSP could cover a much wider area to include a whole continent. Initially, a Patriot battery commander used a surveillance radar unit to detect an incoming ballistic missile. DSP gave commanders information to focus on a particular area. This warning saved much time.

Although Air Force space operators were not trained or equipped originally to handle theater ballistic missile warning, Operation *Desert Storm* allowed them to improve warning time with experience. At the beginning of operations, DSP alerts took five and a half minutes from launch plume detection to notification in the field; by this time the Patriot surveillance radar would have already started to track the ballistic missile. Through training, improved communications and streamlined processes, analysts reduced the delay to only two minutes. This extra time allowed the appropriate Patriot battery to prepare for launches instead of requiring all Patriot batteries to maintain constant surveillance that would reduce combat effectiveness over time. DSP also provided important warning to governments, like Israel, to take defensive measures to protect their populace. This helped dissuade Israel from making counterattacks that might have split the Coalition.

Unfortunately, despite the reduced warning time from DSP, commanders could not use the data sufficiently to destroy SCUD and Al Husayn mobile missile launchers. AFSPC crews could detect the ballistic missile lift off, but sufficient coordination with Coalition aircraft or other forces to destroy the mobile launchers was lacking. Throughout the Persian Gulf War, Iraqi launch crews bedeviled Coalition forces with SCUD launches. Although these launches did not disrupt any significant military activities, they did have some successes. An Iraqi ballistic missile landed on a building in Dhahran, Saudi Arabia, and killed 28 US Army personnel and wounded 97 on February 25, 1991.

The Coalition also required a vast array of reconnaissance information. Providing information to commanders took on a different lens from ones in the Cold War. DMSP provided weather reconnaissance for ground operations and Special Forces. DSP allowed for a surveillance capability of ballistic missile launches. Multi-spectral data from environmental monitoring satellites gave military agencies information about battlefield conditions. The national intelligence agencies took imagery of force concentrations throughout the theater. Under the NRO, space reconnaissance satellites provided a great deal of strategic reconnaissance information. However, usable information required additional support from processing equipment if the satellite's products were transmitted directly to staffs in the theater. Washington sent a data receiver van to the Persian Gulf by December 1990. Satellite operations centers would receive signals from NRO satellites and others could transmit data through DSCS to the theater. This particular van had to manipulate large digital files to give commanders appropriate imagery. A van's crew could handle files at 6.0 megabytes per second. Despite the heavy use of DSCS by many users, it could still process imagery at 1.0 megabytes per second.

Other intelligence agencies used space systems designed for signals and other information gathering. However, DMSP, DSP, LANDSAT, SPOT and aerial platforms like the U-2/TR-1, RC-135 and RF-4 produced much of the tactical information that commanders needed to immediately execute the war.

Current military space applications

Air Force engineers developed space systems under a shroud of secrecy during their infancy. Throughout the Cold War, most Pentagon planners restricted military space capabilities to strategic missions. Technology, strategic focus, resources and organizational attention to more conventional systems limited the expansion of military space. The Persian Gulf War shifted military space systems from the realm of the exotic to the routine. GPS, satellite communications and DSP became everyday tools for Washington to conduct warfare. Space systems evolved from acting in a supplementary role to existing telephonic communications, ground-based radars and celestial navigation to dominating a whole range of activities. During the recent 2003 Iraqi campaign, air forces increased their commercial satellite communications terminals in the region by 560 percent and military satellite communications terminals by 120 percent over pre-conflict levels.

Operations *Enduring Freedom* (OEF) and *Iraqi Freedom* (OIF) forced the US to rely regularly on space systems. Before Washington launched these campaigns, the US military was in the process of converting itself from a force garrisoned largely in Europe and other static positions to an expeditionary force, which requires lighter forces that the Pentagon can ship overseas quickly. Air-transportable units need the same or increased firepower with a smaller support base. Military planners can use many support activities in the continental United States to replace many support activities once stationed with front-line troops.

Rapid communications and information from space-based systems can provide combatant commanders instant data that many parties can share and manipulate, and that can guide these commanders. Additionally, constrained military resources have forced commanders to explore new ways to improve warfighting capabilities. Combined use of joint air, land, sea, and space forces could expand efficiencies and effectiveness. If ground forces can download space imagery on targets, then they can communicate with available aircraft to bomb it or use a nearby surface ship to launch a cruise missile. A single soldier in Afghanistan can use satellite communications to direct a B-2 bomber to deliver a JDAM on a terrorist stronghold in the mountains. JDAM has an inertial navigational system that uses GPS to attack in all weather and at high altitude. On-demand support requested from the user to aircrews, or direct-fire artillery units, also eliminates a layer of fire-support coordination by higher headquarters. Ground force planners who used to organize artillery fire-support missions or close air support can now use a host of military capabilities. This ability is only possible with the sharing of common, instant and reliable information delivered by space systems.

Unlike the pre-1991 Persian Gulf War era, engineers routinely design weapons platforms and systems around space-based services. Unlike Operation *Desert Storm*, where most aircrews had to retrofit their aircraft with GPS receivers, OIF aircrews considered GPS navigation and weapons delivery common practice. Coalition OIF aircrews dropped about 20,000 guided munitions, 32.7 percent of which were JDAMs. Critics might contend higher level commands, with improved communications and precision, might micro-manage a local commander's activities or approve risky operations that they would never have accepted before. Taking actions against a terrorist or insurgent near a religious or other sensitive target, despite greatly improved precision, is still an uncertain proposition. Accidental collateral damage can create tremendous effects. Tactical actions can have strategic impacts given the instantaneous transmission of global news stories from the same space-based communications systems. Intensive command and control requires more, better and faster information which only space systems can deliver.

During the Cold War, the military race for space turned hot. American and Soviet space systems competed for position and capability in orbit. This Soviet Cosmos communications satellite tried to reproduce the same function as DSCS. (DOD)

DSP has been a key element for ballistic missile early warning since its first deployment in November 1970. DSP information was critical for tactical ballistic missile warning during the 1991 Persian Gulf War. This rare DOD mission by the Space Shuttle deploys a DSP satellite. (USAF)

Operations in Afghanistan, Iraq and other locations in the future may take place in areas where there is little infrastructure to support the military. Space systems have allowed expanded operations without the use of conventional support facilities in theater for communications, navigation, and other activities over hostile terrain. Instant communications and other services are particularly vital once a conflict moves from a conventional fight to an insurgency situation. This type of conflict requires vast amounts of information on situations in a constant state of flux. Military, political, intergovernmental and non-governmental organizations need support on an unprecedented scale both in theater and externally to provide activities from conducting bomb sorties to humanitarian aid. Destroyed or nonexistent infrastructure impeded progress during the conflict in Iraq and only space systems allowed for this level of communication.

The same space-based capabilities developed for military purposes are also available to Iraqi and Afghani insurgents. Disposable cell phones, commercial navigational signal receivers, imagery from a host of civilian sources and the ability to establish geographically separated organizations have allowed insurgents to apply space-based systems to the conflict.

Foreign military space systems

Many military space functions that were only obtainable through massive development and production efforts are now purchasable through commercial satellites. Almost any nation that wants to develop a communications network can purchase a satellite and contract with a commercial launch service, which removes the need for nations to operate an expensive space launch complex or fund an aerospace industry. More countries than ever have tried to seek ways to access space. The growing importance of space-based communications, intelligence-gathering, early warning and other systems has also led to many nations developing military space systems to protect their access to these vital resources.

Man-made space objects in orbit (May 2005)				
Operator	**Satellites**	**Space probes**	**Debris**	**Total**
Russia	1,358	35	2,672	4,065
US	920	54	2,977	3,951
PRC	47	0	305	352
France	42	0	294	336
Japan	86	7	54	147
India	30	0	107	137
European Space Agency	34	5	32	71
International Telecom Satellite Organization	61	0	0	61
Globalstar	52	0	0	52
Orbcomm	35	0	0	35
European Telecom Satellite Organization	26	0	0	26
Canada	22	0	1	23
Germany	22	0	1	23
UK	22	0	1	23
Sea Launch	1	0	13	14
Italy	11	0	2	13
Luxembourg	13	0	0	13
Australia	9	0	2	11
Brazil	10	0	0	10
International Maritime Satellite Organization	10	0	0	10
Sweden	10	0	0	10
Argentina	9	0	0	9
Indonesia	9	0	0	9
NATO	8	0	0	8
South Korea	8	0	0	8
Spain	8	0	0	8
Arab Satellite Communications Organization	7	0	0	7
Mexico	6	0	0	6
Saudi Arabia	6	0	0	6
Czech Republic	5	0	0	5
Israel	5	0	0	5
Netherlands	5	0	0	5
Turkey	5	0	0	5
Other	42	3	0	45
Total	2,942	106	6,461	9,509

Source: *Air Force Magazine*

One of the constraints to expanding space operations has always been the cost of putting a payload into orbit. After the *Challenger* disaster, the Air Force invested resources into developing the EELV concept of acquiring a more reliable, cheaper source of space boosters. The Air Force chose the Delta and Atlas family of boosters. (USAF)

The Russian Federation operates an extensive military space program, albeit on a reduced scale from its Soviet Cold War levels, and they still maintain a fairly robust system of military satellites to support their dwindling conventional and strategic nuclear forces. A more interesting case is the PRC, whose military, civil and commercial space programs have advanced significantly since the early 1970s. They have used their ballistic missile development programs to support a military space program, like the US and Russia before them.

While still in its infancy, the PRC has been able to develop a growing military space capability that can be characterized as aiming to counter US space capabilities and deny access to existing systems. The Chinese leadership was impressed by the space capabilities demonstrated by the United States in the 1991 Persian Gulf War and Operation *Allied Force*, the 78-day air campaign over Kosovo from March to June 1999. The Iraqi military was armed, trained and organized like the Chinese People's Liberation Army (PLA) and the PLA leaders realized that they had to modernize their forces to improve their ability to fight in an era where a foe could bring precision and speed to the battlefield. This view has been strengthened by recent US activities in Iraq and Afghanistan. China could now seriously affect the US military's capability by exploiting American dependence on space systems. Washington's superiority in information technology could become an Achilles heel if space systems were unavailable. Although the Chinese are duplicating many American space capabilities, they are also focusing on ASAT functions to deny vital functions to any expeditionary force intervening in a host of areas that could conflict with the PRC's national interests, like Taiwan. Attaining space superiority to guarantee free access to essential space systems is now a key issue. Aside from counterspace capabilities, the PRC leadership also sees the ability to launch space systems as vital for their economic development.

American, European Space Agency (ESA) and Russian space launch capabilities are limited. American space launch services face a backlog of commercial, civil and military satellite launches. After the *Challenger* disaster, a conscious policy decision by Washington to eliminate Space Shuttle commercial space launches resulted in additional pressure on satellite users to find alternate means to put their payloads into orbit. As the ESA's Ariane launch booster and services are also limited and US laws prohibited any American firm from contracting with the then Soviet Union, American commercial satellite manufacturers and users sought other sources, like those offered by an expanding China. Washington believed that Beijing was a counterweight to the Soviets and increased business contacts were crucial to this relationship. China has used its space boosters to launch other nations' satellites, gaining hard currency and technological expertise.

The PRC also had a key national interest to advance space—international prestige. The Chinese launched a manned space mission in 2003 and this technological great leap forward demonstrated the Chinese government's ability to conduct operations, albeit at a limited scope, similar to those undertaken by the US and the Soviet Union in the 1960s.

Chinese military space efforts were aided greatly by American-educated Qian Xuesen who was deported to China in 1955. Qian was accused of espionage, but had worked on a number of American ballistic missile developments including Titan. He became the scientific catalyst for the PRC's ballistic missile and space programs. The PRC first used Soviet R-1s to test ballistic missile concepts. These copies of German V-2s provided a start for Qian and Chinese scientists and engineers built successive Dong Feng (East Wind) ballistic missiles with the latest missile, DF-5, capable of striking the continental United States. The DF family allowed Beijing to expand its reach to space.

US military space systems in orbit (1958–2005)	
Mission	
Communications	126
Weather	44
Navigation	94
Test vehicle	3
Other	83
Total mission	350
ASAT	
ASAT interceptors	33
ASAT targets	2
SDI test vehicles	11
Total ASAT	46
Reconnaissance	
Imagery	250
Electronic intercept	49
Ocean surveillance	46
Nuclear detection	12
Radar calibration	40
Early warning	39
Total reconnaissance	436
TOTAL	**832**

Source: *Air Force Magazine*

The Chinese government has attempted over 76 orbital launches and successfully placed more than 47 space systems into orbit. By comparison, the US has made 1,284 launches placing 1,773 satellites into orbit through to 2003; Russian space personnel have attempted 2,628 orbital launches. The first Chinese spacecraft launched into orbit was its 172kg Dong Fang Hong-1 (DFH-1) on April 24, 1970. This satellite broadcast a recording of *The East is Red* for 26 days. Chinese launch services used a modified DF-4 that Beijing rechristened as the Chang Zheng (CZ) or Long March. Since the successful launch of the DFH-1, the Chinese space program has expanded both its SLV and satellite capability. The Chinese have four series of the Long March, CZ-1 to CZ-4. Each succeeding series also has versions that vary based on third stage, improved first and second stages, strap-on solid rocket boosters and other modifications. These SLVs allow the PRC to launch small LEO satellites to larger GEO systems into orbits.

Since the 1970 launch of the DFH-1, Beijing has deployed 47 orbiting satellites, including communications, environmental monitoring, photoreconnaissance, weather and navigational systems for commercial, civil, and military purposes. The PLA can rely on several military systems that include a photoreconnaissance, two commercial/military communications, one military purpose communications, two navigational and a maritime surveillance satellite. The two Chinese navigational satellites provide a very limited capability; however, the PRC has expanded its participation in the European Galileo navigational satellite system to reduce its dependence on GPS.

Chinese military space capability is growing, but commercial demand may outstrip current and future systems. Economic development and the rising standard of living among the populace may focus commercial and civil space programs towards the exploding market for cell phones, weather, remote sensing, data and other non-military applications. The PRC has turned to building its own expanding commercial space capabilities. One controversial aspect of commercial space launch capability is the sharing of technology. The PRC's CZ launch reliability suffered during the 1990s and, allegedly, American aerospace firms provided technology to improve launch reliability and guidance capabilities that could be used to improve PRC ballistic missiles.

Commercial space development has also allowed shared information from France, Germany and Brazil. The PRC and Brazil have developed joint remote-sensing satellites that could support both civilian and military requirements. China's Ziyuan-1, launched in October 2003, has a resolution of 19m and its Ziyuan-2 successor has worldwide coverage that can deliver digital imagery.

RIGHT **Future space threats**
A nation could attempt to disable another nation's space assets by a variety of methods. Some countries can build smaller, cheaper micro-satellites that could send nano-satellites, small rocket, or other devices to destroy components of a targeted satellite. Here small nano-satellites attach to and then explode on a commercial communications satellite. Military forces, like the United States, frequently lease commercial satellite circuits to supplement their own military satellite capabilities.

These contacts have allowed the PRC to enhance their launch readiness, improve satellite capabilities and expand political influence. The PRC is also developing smaller boosters to launch tactical satellites during a military conflict.

The PRC has an extensive space launch support system that includes three launch complexes and a space tracking system that spans ground stations in the Atlantic, Africa, South Asia, the South Pacific and China itself; it also has two dedicated space control centers. Despite a growing SLV and satellite industry, the PRC cannot match global United States space capability at this time. Instead, the PRC may try to build a counterspace capability.

The PRC sees Washington's construction of a limited missile defense system as a threat to its strategic ballistic missile retaliatory capability. Beijing fears that if the missile defense system is large enough to defeat a retaliatory strategic nuclear strike, then Washington could launch a preemptive strike with impunity. China's ability to ensure its perceived national survival by maintaining a credible nuclear strike capability is vital. Defeating key early warning and tracking satellites could make a missile defense system ineffective.

Similarly, other activities to eliminate American space capabilities could delay or deny capabilities to launch a large expeditionary force to counter any PRC military movement to take over Taiwan. Despite military basing in Guam, Okinawa, Japan, and Korea, Washington would need to deploy military forces far from logistical and support forces in an engagement in Asia. Like its deployments in Iraq and Afghanistan, it would rely on space systems for operations.

China has several options to create an ASAT capability. They range from a simple computer virus attack to micro-satellites. Some counterspace activities seem less likely than others. The USAF has based many of the key space ground centers, such as satellite operations centers, in the continental US and a conventional assault on these positions would be difficult considering heightened security against terrorists and geographic distance. Similarly, ASAT attacks on American satellites using an EMP attack via a nuclear explosion (like Program 437) would not only damage American, but Chinese and other nations' satellites.

Most likely, Beijing and the PLA leadership would use systems limiting collateral damage and specifically target particular American military satellites. Aside from computer network attacks, Beijing could try to jam signals between space systems and ground stations or receivers. Iraq tried to jam GPS signals during Operation *Iraqi Freedom* in 2003 and failed; however, the PRC's advancing information technology industry may allow it to find effective means to jam GPS and other signals. DSCS and other national communications systems offer jam-resistant capabilities; however, leased communications satellites may not.

China has also experimented with DEWs, along with the US and Russia. The US successfully tested a Mid-Infrared Advanced Chemical Laser for ballistic missile defense and it has plans to deploy an airborne laser. Russia has allegedly used low-powered lasers to blind American satellites. China could also build ground-based DEWs powered by a nuclear power plant to strike or disrupt LEO satellites. These weapons only have to disable components, like optical systems. The Chinese military could also use high-powered microwaves to disrupt or disable satellite operations.

Advanced technology has allowed miniaturization of components that has reduced satellites' weight and increased capacity. If Beijing wants to create a single-purpose, limited-duration spacecraft, then it could create a fleet of small

Future space threats

Russian military/civil space systems in orbit (1958–2004)	
Force support	
Communications	869
Weather	75
Navigation	236
Geodesy	34
Earth sciences	312
Other	178
Total force support	1,704
Weapon	
ASAT interceptor test	20
ASAT targets	18
FOBS	18
Total weapon	56
Reconnaissance	
Imagery	809
Electronic intercept	133
Oceanic intelligence	85
Early warning	83
Total reconnaissance	1,110
TOTAL	**2,870**

Source: *Air Force Magazine*

micro-satellites whose sole purpose is an ASAT one. The Chinese are building several micro-satellites, weighing from 10 to 100kg, which can perform remote-sensing functions including electro-optical and radar imaging capabilities with a 50m resolution. The US is trying to build similar tactical satellites capable of providing combatant commanders with temporary space-based capabilities. The Chinese orbited its Tsinghua-1 micro-satellite on a Russian booster in June 2000. The Tsinghua-1 had a GPS receiver and multi-spectral sensor. The PRC could deploy these types of micro-satellites rapidly with solid fuel ballistic missiles. Its latest ICBM, the DF-31, is a road-mobile system. Similar launch capabilities for micro-satellites or smaller systems could allow the PRC to launch a space attack.

The PRC's scientific and space technology communities have also examined the use of a new nano-class of satellites weighing less than 10kg. If the Chinese military could deploy a large number of nano-satellites, it could conduct a widescale attack. Satellite controllers could move them towards a target space system to disrupt it by jamming, disabling a solar panel or other such scheme. These satellites could also act as an insurance policy for the PRC as "parasites" that could stay near its target and conduct an attack on demand if Beijing suspects an attack or wants to conduct military operations. Disabling or destroying American space capabilities might allow the Chinese to gain valuable time to secure their objective or prepare for an attack.

The PRC is not the only country to have taken notice of America's space reliance. North Korea, Iran and other authoritarian nations that have access to ballistic missiles and a nuclear weapon could replicate the USAF's Thor nuclear-based ASAT. The North Korean government may not care about destroying spacecraft since they may not be as dependent on these systems nor have an investment in them. An attack could also provide a demonstration of their will to use nuclear weapons and challenge an American response. Washington may debate whether destroyed satellites might constitute a reason to go to war, despite EMP effects on the surface that might cause casualties from disruption of electrical grids and electronic component failures.

North Korea or Iran could conduct an EMP attack on space systems during the opening stages of a conflict. To blind imagery, communications, and navigational capabilities, crews could launch a ballistic missile armed with a small nuclear warhead to an altitude sufficient to detonate the warhead. This high-altitude explosion would provide a catastrophic impact on American deployment of forces and degrade surface systems. Such an attack could render unhardened electronic systems from Seoul to Tokyo useless. Economic and social chaos would ensue and allow North Korean forces to conduct combat operations with a massive diversion.

America's space future

The US has unlimited space capabilities compared to any potential rivals. Although countries like Russia or China may not compete with building as many or sophisticated space systems as the US, they could attempt to deny space access to American military forces. One means to make certain that the US can guarantee its use of space is to ensure that the nation and her allies have space supremacy. If the US can maintain its ability to launch, orbit and use its satellites, while denying the same capability to a rival, then it can maintain its ability to conduct a wide range of global operations. New threats to not only military but also commercial satellites need a solution. Developing passive defenses could give a means of protection from natural and man-made threats to systems. Unfortunately, commercial satellite firms might balk at the cost and weight to protect these assets. How the US protects these systems without appearing to threaten foreign ones is a challenge.

The US can develop and deploy many enhanced space capabilities that can replace many terrestrial systems. Real-time radar systems can ensure instant battlefield surveillance that can improve greatly command and control efforts. Similarly, communications and information-sharing devices can enable individual soldiers to conduct operations never envisioned before. Nations can use space-based systems to target and attack locations and forces with unparalleled accuracy. For the US to attain these capabilities, it must have space supremacy.

The US military could gain space supremacy through a number of ways. It can use an active ASAT program that can identify, track and eliminate hostile spacecraft. The nation might create a rapid deployment system that could replace, albeit with limited functions and duration, or increase particular space-based capabilities. Washington could also deploy conventional forces that could disrupt or deny any space threat via an attack on ground stations and launch complexes. A combination of these programs would offer a wealth of options to a commander. Today, military commanders still rely on space systems to avoid a "nuclear" Pearl Harbor, but now must also consider avoiding a surprise attack on its space systems a priority.

A return to Program 437 capabilities or a kinetic kill vehicle like the modified F-15 is highly doubtful. Collateral damage caused by a nuclear detonation, such as EMP, radiation and X-rays, would destroy the very systems the US seeks to protect. Debris created from an MHV would create a debris field. DEWs on board an orbiting satellite have fuel limitations, unless they were close enough and had sufficient precision to target an offending vital component. Electronic jamming, spoofing, or using a similar means to temporarily disable a space system is available today. Although this conventional weapon could succeed, could the US ensure that it could disrupt every satellite, or worse, enemy micro- or nano-satellites devised as ASATs?

If the US achieves control of space, it has other options that it could exercise. This includes the potential placement of weapons in space to counter ASATs, provide ballistic missile defense, or deploy weapons from space to the Earth's surface. Like fortresses in the past, where military units protected geographic locations, people and assets, these sites also served as starting points to launch campaigns. In this case, space offers some unique characteristics to employ such as the maneuvering spacecraft that could provide a range of options for military commanders. One of the key limitations for space operations has been a lack of a rapid delivery vehicle to deploy spacecraft. The Space Shuttle was supposed to

Demand for space-based assets not only includes communications, but weather and other functions. This Geostationary Operational Environmental Satellite spacecraft collects weather data used in forecasting. National Oceanic and Atmospheric Administration personnel operate the satellites, which were first launched on October 16, 1975. (NOAA)

give the nation a relatively fast turnaround to put satellites into LEO. However, shifting military launch requirements from the Space Shuttle to expendable boosters caused the nation to revamp its launch capabilities. Launch crews still require weeks to prepare an expendable SLV. Tactical satellites might not have the payload or duration for many missions. A manned hypersonic spacecraft that could give speeds of Mach 10 or more, like the 1960-era X-20 Dyna-Soar, could allow Washington to conduct many operations in a more operational manner, including potential ASAT and weapons deployment. The Air Force has experimented, for years, on developing a hypersonic vehicle. One proposed vehicle could carry up to 454kg from high altitude or, potentially, from LEO to a point halfway around the world in 45 minutes. One possible weapon may use hypervelocity rods that use kinetic energy to destroy bunkers or other targets: titanium rods, "Rods from God," weighing about 113kg and traveling at speeds of over 11,500kmph could deliver a devastating attack, assuming they were precisely delivered on a target. A military space transport could also provide rapid deployment of replacement satellites or fielding of tactical satellites.

If the United States places weapons, like ASATs, or a platform to deliver a weapon from space, then it faces several challenges. Despite the well-intentioned aims of deploying space weapons to protect valuable satellites, international issues could arise. Other nations might see this as a threat and develop similar weapons, starting an arms race. Cost and technical, political, diplomatic and legal considerations are considerable. For example, although the placement of weapons seems like a violation of international law, there is no prohibition of the stationing of non-nuclear weapons in space according to the 1967 Outer Space Treaty. Current concepts of space-based weapons in space do not include nuclear weapons. The United States has also withdrawn from the 1972 Antiballistic Missile Treaty that would have made space-based missile defense systems illegal. However, other nations may not adhere to such limitations; nuclear weapons could be placed in space in response to a hurried program to ensure a country has a counterweight to the United States.

Like the X-15, this hypersonic test X-43 demonstrates the feasibility of launching a high-speed vehicle that could enter space. The US could develop a military space vehicle in the near future. (USAF)

Space weaponry provides several key advantages for nations. These systems can improve and support air, land, sea, information, and space operations against a host of threats. Attacking from such a high altitude also allows a country to have a global strike capability without having to maneuver forces into position or rely on base access from other nations. Certain global operations might not require "boots on the ground," rather a quick strike on a military or economic target. Depending on the weapons used and the level of attack, a space weapons program can deliver a devastating punch; for example destroying ballistic missile threats before they are launched or during their vulnerable boost phase. If the space weapons are used against a developing country, then they might not have any means to defend against these weapons.

Space weapons also have some major obstacles to address. Other than the cost and technical, political, and legal obstacles, there are several operational concerns. Orbiting an unmanned space weapons platform, like a fortress, creates a very predictable flight path. A potential rival can assign a micro- or nano-satellite to attack the platform on orbit, or the foe could use a number of ASAT devices to disable it given precise orbital information. Once the space weapons platform is in orbit, it may have little flexibility to maneuver given limited fuel stored on board spacecraft; in many ways they would become fixed fortifications. Resupply of munitions and fuel also becomes a considerable issue in space. Without a means to replenish kinetic weapons and fuel, or make repairs, these platforms may be a one-use only system.

Undoubtedly, space weapons are a definite possibility in the future. Will countries consider them a new threat to the world, or will they simply become an extension of the current aerial arsenal? Military officers once considered ballistic missiles as mere extensions of field artillery, yet today they transit space as a routine part of their flight path. Such weapons of mass destruction have proliferated with the spread of information and the desire for nations to possess them. Given the ability of the US, Russia and China to produce space weapons today, it is a possibility that a similar situation may develop in the future.

The USAF is testing and developing an airborne laser for ballistic missile defense. Future systems could also evolve into an ASAT carrier. The technology to track a ballistic missile and direct and fire a laser at it shares many of the same characteristics needed for an ASAT weapon. (USAF)

US military space sites

There are several military space sites located within the US and overseas. Military security officials have closed most of these sites to the public, but some do offer visitors a chance to view some unique aspects of space operations. Additionally, several museums display a glimpse of military and civilian space activities from the past.

The USAF's two major launch sites, Vandenberg and Cape Canaveral, announce unclassified space and test missile launches. Viewers can get a chance to see a spectacular demonstration of space power, especially during a night launch.

Vandenberg is located on the central California coast and is 13km northwest of Lompoc. Home of the 14th Space Wing, Vandenberg is one of the largest Air Force installations at 40,000 hectares. A visitor can easily reach Vandenberg by following US Highway 101 south from San Francisco or north from Los Angeles. From either direction coming via US Highway 101, take Highway 1 and then State Road 20 that will deliver you to the front gate of Vandenberg. Base tours are available if one contacts the Public Affairs office beforehand on (+1) 805 606 3595. Air Force personnel offer a public tour on the second and fourth Wednesday morning of each month at 1000hrs. You need to contact the Public Affairs office at least two weeks before the desired date and visitors must be at least ten years old. Visitors must show two forms of photo identification and get approval from Public Affairs before proceeding on the tour, which is free.

Cape Canaveral AFS, located on the Florida coast, serves as the launch site for satellites placed into equatorial orbits. This site, as well as the Kennedy Space Center, has been the subject of much publicity concerning manned and unmanned space activities. These launch facilities continue today and will flourish into the foreseeable future. (USAF)

Cape Canaveral is located on the Atlantic side of the Florida coast on a peninsula. Cape Canaveral is adjacent to NASA's Kennedy Space Center, which offers many amenities including a visitor's center and museum of civilian space launches. From Orlando go east on State 528 to reach the site; from Miami take US Interstate 95 north to reach Cape Canaveral, which contains the Air Force Space and Missile Museum that hold static displays, ballistic missiles, rockets and other space-related equipment; admission is free. The museum is open daily and for more information contact the curator on (+1) 321 853 9171.

An interesting aspect of space operations for a visitor is to view the Peterson AFB area near Colorado Springs, Colorado. Peterson contains the AFSPC headquarters and is on the eastern edge of the city. Located here is a museum offering static display aircraft and missiles that have served in continental air defense and space missions. The museum is open from Tuesday to Saturday, 0830 to 1630hrs. Due to security restrictions, visitors without a a valid military or Department of Defense identification card must contact the museum to receive permission to visit. Current requirements for a site visit are available by calling (+1) 719 556 4915. Additionally, Cheyenne Mountain Air Force Station, southwest of the city, contains NORAD, the nerve center for missile warning and space detection. The Cheyenne Mountain complex is a unique site built by the Air Force under the Rocky Mountains to be secure from a nuclear strike. Although the USAF has conducted public tours of the facility, they are infrequent due to changing security requirements.

Interested persons can also visit the National Museum of the United States Air Force at Wright Patterson AFB, Ohio, near the city of Dayton. Displays include rocket, missile and space artifacts. The museum is open daily from 0900 to 1700hrs, except on Thanksgiving, Christmas and New Year's Day. The museum is easily accessible from Interstates 70, 75 or 675 and admission is free. Part of the museum, which includes presidential aircraft and research and development displays, is located on an active portion of Wright Patterson AFB and a form of government-issued photo identification or passport is needed for visitors to see this area. The museum can be contacted on (+1) 937 255 3284 for more information.

The National Air and Space Museum has two locations: the main museum is on the National Mall in Washington, DC. It contains several space exhibits and is the most visited museum of all of the Smithsonian Institution facilities. The second museum, the Steven F. Udvar-Hazy Center is located near Dulles International Airport in nearby Chantilly, Virginia, at the intersection of State Routes 28 (Sully Road) and 50. Admission to both museums is free and they are open daily from 1030 to 1730hrs, except on Christmas Day.

Cape Canaveral has been a keystone of the American civil, commercial and military space programs since the late 1950s. This early photograph of Cape Canaveral illustrates the extensive launch pad activity that dominated the site. (USAF)

Bibliography

Barter, Neville, ed. *TRW Space Data* (Redondo Beach, CA: TRW Space & Technology Group, undated)

Chun, Clayton K.S. *Shooting Down a "Star": Program 437, the US Nuclear ASAT System and Present-Fay Copycat Killers* (Maxwell AFB, AL: Air University Press, 2000)

Collins, John M. *Military Space Forces* (London: Pergamon-Brassey's, 1989)

Day, Dwayne A., John M. Logsdon, and Brian Latrell, eds. *Eye in the Sky: The Story of the Corona Spy Satellites* (Washington, DC: Smithsonian Institution Press, 1998)

DeSciscíolo, Dominic "China's Space Development and Nuclear Strategy" ed. Lyle J. Goldstein *China's Nuclear Force Modernization* (Newport, RI: US Naval War College Press, 2005)

Gonzales, Daniel *The Changing Role of the U.S. Military in Space* (Santa Monica, CA: RAND Corporation, 1999)

Hall, R. Cargill and Jacob Neufeld *The U.S. Air Force in Space 1945 to the 21st Century* (Washington, DC: Air Force History and Museums Program, 1998)

Hobbs, David *Space Warfare: 'Star Wars" Technology Diagrammed and Explained* (New York: Prentice Hall Press, 1986)

Keaney, Thomas A. and Eliot A. Cohen *Revolution in Warfare?* (Annapolis, MD: Naval Institute Press, 1995)

Lambeth, Benjamin S. *Mastering the Ultimate High Ground* (Santa Monica, CA: RAND Corporation, 2003)

McDougall, Walter A. *The Heavens and the Earth* (Baltimore, MD: The Johns Hopkins University Press, 1985)

Muolo, Michael J. *Space Handbook: A War Fighter's Guide to Space* (Maxwell AFB, AL: Air University Press, 1993)

Office of the Secretary of Defense *Annual Report on the Military Power of the People's Republic of China* (Washington, DC: Department of Defense, various years)

__________, *Conduct of the Persian Gulf War* (Washington, DC: Department of Defense, April 1992)

Peebles, Curtis *The Corona Project: America's First Spy Satellites* (Annapolis, MD: Naval Institute Press, 1997)

__________, *High Frontier* (Washington, DC: Air Force History and Museums Program, 1997)

Preston, Bob, et al. *Space Weapons, Earth Wars* (Santa Monica, CA: RAND Corporation, 2002)

Richelson, Jeffrey T. *America's Space Sentinels: DSPO Satellites and National Security* (Lawrence, KS: University of Kansas Press, 1999)

Ruffner, Kevin C., ed. *Corona: America's First Satellite Program* (Washington, DC: Central Intelligence Agency, 1995)

Sellers, Jerry Jon *Understanding Space: An Introduction to Astronautics* (New York: McGraw-Hill, 1994)

"Space Almanac" *Air Force Magazine* (several years, August issue).

Spires, David N. *Beyond Horizons: A Half Century of Air Force Space Leadership* (Maxwell AFB, AL: Air University Press, 1998)

Stares, Paul B. *The Militarization of Space U.S. Policy, 1945–1984* (Ithaca, NY: Cornell University Press, 1985)

Waldron, Harry N. *50 Years Air Force Space & Missiles* (Los Angeles, CA: Office of History Space and Missile Systems Center, 2004)

Glossary

ABM	anti-ballistic missile
ADC	Aerospace Defense Command
AEHF	Advanced Extremely High Frequency Satellite Communications System
AFB	Air Force Base
AFS	Air Force Station
AFSATCOM	Air Force Satellite Communications
AFSPC	Air Force Space Command
ARDC	Air Research and Development Command
ASAT	anti-satellite
ATBM	anti-theater ballistic missile
ATO	air tasking order
BMEWS	Ballistic Missile Early Warning System
CENTCOM	Central Command
CEP	circular error probable
CIA	Central Intelligence Agency
CZ	Chang Zheng
DEW	directed energy weapon
DF	Dong Feng
DFH	Dong Fang Hong
DMSP	Defense Meteorological Support Program
DOD	Department of Defense
DSCS	Defense Satellite Communications System
DSP	Defense Support Program
EELV	Evolved Expendable Launch Vehicle
EMP	electromagnetic pulse
ESA	European Space Agency
FLTSATCOM	Fleet Satellite Communications System
FOBS	fractional orbiting bombardment system
GBS	Global Broadcast System
GEO	geosynchronous Earth orbit
GPS	Global Positioning System
GTO	geosynchronous transfer orbit
HEO	high Earth orbit
ICBM	intercontinental ballistic missile
IDCSP	Initial Defense Communications Satellite Program
IR	infrared
IRBM	intermediate range ballistic missile
JDAM	Joint Direct Attack Munition
LEO	low Earth orbit
MEO	medium Earth orbit
MHV	miniature homing vehicle
MIDAS	Missile Detection Alarm System
MILSATCOM	Military Satellite Communications
MILSTAR	Military Satellite Communications System
MOBS	Multiple Orbiting Bombardment System
NASA	National Aeronautics and Space Administration
NGA	National Geospatial-Intelligence Agency
NOAA	National Oceanic and Atmospheric Administration
NORAD	North American Aerospace Defense Command
NRO	National Reconnaissance Office
NSA	National Security Agency
OEF	Operation *Enduring Freedom*
OIF	Operation *Iraqi Freedom*
PLA	People's Liberation Army
PRC	People's Republic of China
RV	reentry vehicle
SAC	Strategic Air Command
SAMOS	Satellite and Missile Observation Systems
SDI	Strategic Defense Initiative
SDS	Satellite Data System
SLBM	submarine-launched ballistic missile
SLV	space launch vehicle
TERS	Tactical Event Reporting System
UFO	UHF Follow-On Satellite
USAF	United States Air Force
WDD	Western Development Division
WS	weapons system

Index

Figures in bold refer to illustrations.